Iosias Jody (Ed.)

Grindelwald Railway Station

Iosias Jody (Ed.)

Grindelwald Railway Station

Grindelwald, Bern, Grindelwald Grund railway station, Schwendi railway station, Wengernalpbahn

Cred Press

Imprint

Publisher:
Cred Press is a trademark of
International Book Market Service Ltd., 17 Rue Meldrum, Beau Bassin, 1713-01 Mauritius
Email: info@bookmarketservice.com
Website: www.bookmarketservice.com

Published in 2011

Printed in: U.S.A., U.K., Germany. This book was not produced in Mauritius.

ISBN: 978-613-5-79955-2

Contents

Articles

References

Article Licenses

Grindelwald railway station

Grindelwald regional rail	
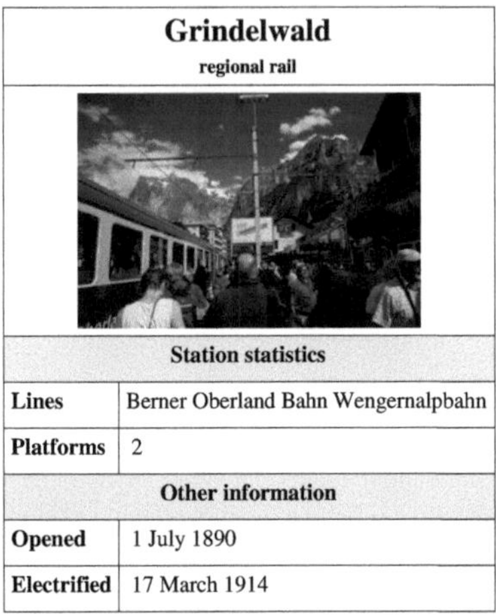	
Station statistics	
Lines	Berner Oberland Bahn Wengernalpbahn
Platforms	2
Other information	
Opened	1 July 1890
Electrified	17 March 1914

Grindelwald railway station serves the village of Grindelwald in the Swiss canton of Bern.

Next station East: **Terminus**	Berner Oberland Bahn	Next station West: **Schwendi**

Next station East: **Grindelwald Grund**	**Wengernalpbahn**	Next station West: **Terminus**

Grindelwald

Grindelwald	
 Grindelwald and Wetterhorn	
Country Switzerland **Canton** Bern **District** Interlaken-Oberhasli 46°37′N 8°02′E 	
Population	3860 (Dec 2009)[1]
- Density	23 /km^2 (58 /sq mi)
Area	171.1 km^2 (66.1 sq mi)
Elevation	1034 m (3392 ft)
Postal code	3818
SFOS number	0576
Mayor	Emanuel Schläppi
Localities	Alpiglen, Grund, Itramen, Mühlebach, Schwendi, Tschingelberg, Wargistal
Surrounded by	Brienz, Brienzwiler, Fieschertal (VS), Guttannen, Innertkirchen, Iseltwald, Lauterbrunnen, Lütschental, Meiringen, Schattenhalb
Twin towns	Azumi, now Matsumoto (Japan)
Website	www.grindelwald.com [2] SFSO statistics [3]

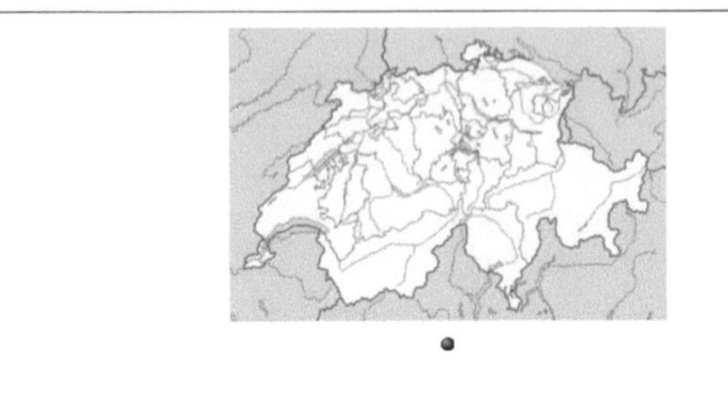

Grindelwald

View map of Grindelwald

Grindelwald is a municipality in the Interlaken-Oberhasli administrative district in the canton of Bern in Switzerland. The village is located at 1034 meters (3392 ft) above sea level in the Bernese Alps.

Winter sports

Main street

Long famed as a winter tourist destination with slopes for beginners, intermediates and the challenges of the Eiger glacier for the experienced, there are activities for the non-skiers, from tobogganing to groomed winter hiking tracks. It is the usual starting point for ascents of the Eiger and the Wetterhorn. Nowadays Grindelwald is also a popular summer activity resort with many miles of hiking trails across the Alps.

Geography

Grindelwald has an area, as of 2009, of 171.08 km^2 (66.05 sq mi). Of this area, 49.47 km^2 (19.10 sq mi) or 28.9% is used for agricultural purposes, while 28.02 km^2 (10.82 sq mi) or 16.4% is forested. Of the rest of the land, 3.06 km^2 (1.18 sq mi) or 1.8% is settled (buildings or roads), 1.37 km^2 (0.53 sq mi) or 0.8% is either rivers or lakes and 89.21 km^2 (34.44 sq mi) or 52.1% is unproductive land.[4]

Of the built up area, housing and buildings made up 1.0% and transportation infrastructure made up 0.6%. 12.9% of the total land area is heavily forested and 2.3% is covered with orchards or small clusters of trees. Of the agricultural land, 5.1% is pastures and 23.8% is used for alpine pastures. All the water in the municipality is in rivers and streams. Of the unproductive areas, 6.6% is unproductive vegetation, 24.0% is too rocky for vegetation and 21.6% of the land is covered by glaciers.[4]

The municipality is quite large and is divided into seven mountain communities. However the municipality is dominated by the large village and tourist center of Grindelwald.

History

Grindelwald is first mentioned in 1146 as *Grindelwalt*.[5]

Demographics

Grindelwald area from above

Grindelwald has a population (as of 31 December 2009) of 3860.[1] As of 2007, 15.8% of the population was made up of foreign nationals. Over the last 10 years the population has decreased by 3.4%. Most of the population (as of 2000) speaks German (86.8%), with Portuguese being second most common (4.5%) and French being third (1.7%).

In the 2007 election the most popular party was the SVP which received 54.8% of the vote. The next three most popular parties were the FDP (14.2%), the Green Party (10.1%) and the SPS (8.6%).

The age distribution of the population (as of 2000) is children and teenagers (0–19 years old) make up 21.1% of the population, while adults (20–64 years old) make up 62.6% and seniors (over 64 years old) make up 16.3%. The entire Swiss population is generally well educated. In Grindelwald about 66.8% of the population (between age 25-64) have completed either non-mandatory upper secondary education or additional higher education (either University or a *Fachhochschule*).

Grindelwald has an unemployment rate of 2.66%. As of 2005, there were 365 people employed in the primary economic sector and about 138 businesses involved in this sector. 378 people are employed in the secondary sector and there are 53 businesses in this sector. 1,961 people are employed in the tertiary sector, with 225 businesses in this sector.[6]

The historical population is given in the following table:[5]

View from mountain in Grindelwald

Year	Population
1764	1,816
1850	2,924
1900	3,346
1920	2,998
1950	3,053
2000	4,069

Transportation

Railway station

Grindelwald can be reached by train (Berner Oberland Bahn) from Interlaken.

The Wengernalpbahn connects Grindelwald to the Kleine Scheidegg from where the Jungfraubahn ascends inside the Eiger to the Jungfraujoch and trains descend to Wengen.

The Gondelbahn Grindelwald-Männlichen connects Grindelwald with the Männlichen and with onward travel on the Luftseilbahn Wengen-Männlichen offers an alternative route to Wengen.

The minor summit of First is accessible by ski lift from Grindelwald.

In the media

Many scenes of the documentary film The Alps were shot in the region of Grindelwald, particularly on the north face of the Eiger. The James Bond film *On Her Majesty's Secret Service* includes a chase through a skating rink and Christmas festival in Grindelwald.[7] Grindelwald's mountains were used as the basis for the view of Alderaan in *Star Wars Episode III: Revenge of the Sith*.[8]

Some of the action scenes in *The Golden Compass* were also shot in Grindelwald.[9]

Famous residents

- Richard Wagner
- Martina Schild, alpine skier, runner-up in the 2006 Winter Olympics women's downhill race
- Hedy Schlunegger, Olympic champion 1948 in downhill skiing
- Oleg Protopopov and Ludmila Belousova, 1964 and 1968 Olympic figure skating champions

See also

- Swiss Alps

References

[1] Swiss Federal Statistical Office (http://www.bfs.admin.ch/bfs/portal/de/index/themen/01/02/blank/data/01.html), MS Excel document – *Bilanz der ständigen Wohnbevölkerung nach Kantonen, Bezirken und Gemeinden* (German) accessed 25 August 2010
[2] http://www.grindelwald.com
[3] http://www.bfs.admin.ch/bfs/portal/en/index/regionen/regionalportraets/gemeindesuche.html?geographyID=0576&FormEncoding=UTF-8
[4] Swiss Federal Statistical Office-Land Use Statistics (http://www.bfs.admin.ch/bfs/portal/de/index/themen/02/03/blank/data/gemeindedaten.html) 2009 data (German) accessed 25 March 2010
[5] *Grindelwald* in German (http://www.hls-dhs-dss.ch/textes/d/D331.php), French (http://www.hls-dhs-dss.ch/textes/f/F331.php) and Italian (http://www.hls-dhs-dss.ch/textes/i/I331.php) in the online *Historical Dictionary of Switzerland.*
[6] Swiss Federal Statistical Office (http://www.bfs.admin.ch/bfs/portal/en/index/regionen/regionalportraets/gemeindesuche.html) accessed 12-Jun-2009
[7] IMDB-On Her Majesty's Secret Service (http://www.imdb.com/title/tt0064757/)
[8] IMDB-Star Wars Episode III: Revenge of the Sith (http://www.imdb.com/title/tt0121766/locations)
[9] IMDB-The Golden Compass (http://www.imdb.com/title/tt0385752/locations)

External links

- Grindelwald (http://www.grindelwald.com) official website
- Webcams (http://www.swisspanorama.com/)
- *Grindelwald* in German (http://www.hls-dhs-dss.ch/textes/d/D331.php), French (http://www.hls-dhs-dss.ch/textes/f/F331.php) and Italian (http://www.hls-dhs-dss.ch/textes/i/I331.php) in the online *Historical Dictionary of Switzerland.*

Bern

Bern	
 Aerial view of the Old City	
Country Switzerland **Canton** Bern **District** Bern-Mittelland administrative district 46°57′N 7°27′E	
Population	123466 (Dec 2009)[1]
- Density	2393 /km^2 (6197 /sq mi)
Area	51.6 km^2 (19.9 sq mi)
Elevation	542 m (1778 ft)
- Highest	864 m - Gurten, Bern
- Lowest	480 m - Aare
Postal code	3000
SFOS number	0351
Mayor (list)	Alexander Tschäppät SPS/PSS
Demonym	Berner
Surrounded by	Bremgarten bei Bern, Frauenkappelen, Ittigen, Kirchlindach, Köniz, Mühleberg, Muri bei Bern, Neuenegg, Ostermundigen, Wohlen bei Bern, Zollikofen
Website	www.bern.ch [2] SFSO statistics [3]

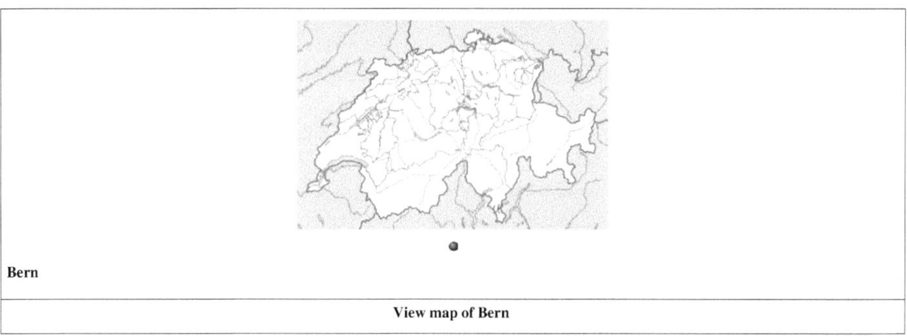

Bern

View map of Bern

The city of **Bern** or **Berne** (German: *Bern*, pronounced [ˈbɛrn] (listen); French: *Berne* [bɛʁn]; Italian: *Berna* Italian pronunciation: [ˈbɛrna]; Romansh: *Berna* [ˈbɛrnə]; Bernese German: *Bärn* [b̥æːrn]) is the *Bundesstadt* (federal city, *de facto* capital) of Switzerland, and, with (as of December 2009) a population of 131,000, the fourth most populous city in Switzerland.[4] The Bern agglomeration, which includes 43 municipalities,[5] has a population of 349,000.[6] The metropolitan area had a population of 660,000 in 2000.[7] Bern is also the capital of the Canton of Bern, the second most populous of Switzerland's cantons.

The official language of Bern is German, but the main spoken language is the Alemannic dialect called Bernese German.

In 1983 the historic old town in the centre of Bern became a UNESCO World Heritage Sites, and Bern is ranked among the world's top ten cities for the best quality of life (2010).[8]

Name

The etymology of the name *Bern* is uncertain. According to the local legend, based on folk etymology, Berchtold V, Duke of Zähringen, the founder of the city of Bern, vowed to name the city after the first animal he met on the hunt, and this turned out to be a bear. It has long been considered likely that the city was named after the Italian city of Verona, which at the time was known as *Bern* in Middle High German. As a result of the find of the Berne zinc tablet in the 1980, it is now more common to assume that the city was named after a pre-existing toponym of celtic origin, possibly **berna* "cleft".[9] The bear was the heraldic animal of the seal and coat of arms of Bern from at least the 1220s. The earliest reference to the keeping of live bears in the *Bärengraben* dates to the 1440s.

History

Early History

The construction of the Untertor-bridge in Bern, Tschachtlanchronik, late 15th century

No archaeological evidence that indicates a settlement on the site of today's city centre prior to the 12th century has been found so far. In antiquity, a celtic *oppidum* stood on the "Engehalbinsel" north of Bern, fortified since the 2nd century BC (late La Tène period), thought to be one of the twelve oppida of the Helvetii mentioned by Caesar. During the Roman era, there was a Gallo-roman *vicus* on the same site. The Bern zinc tablet has the name *Brenodor* "dwelling of Breno". In the Early Middle Ages, there was a settlement in Bümpliz, now a city district of Bern, some 4 km (2 mi) from the medieval city.

The medieval city is a foundation of the Zähringer ruling family, which rose to power in Upper Burgundy in the 12th century. According to 14th century historiography (*Cronica de Berno*, 1309), Bern was founded in 1191 by Berthold V, Duke of Zähringen.

In 1218, after Berthold died without an heir, Bern was made a free imperial city by the Holy Roman Emperor Frederick II.

Old Swiss Confederacy

In 1353 Bern joined the Swiss Confederacy, becoming one of the "eight cantons" of the formative period of 1353 to 1481. Bern was invaded and conquered Aargau in 1415 and Vaud in 1536, as well as other smaller territories, thereby becoming the largest city-state north of the Alps, by the 18th century comprising most of what is today the canton of Berne and the canton of Vaud.

Modern history

Bern in 1638

The city grew out towards the west of the boundaries of the peninsula formed by the river Aare. Initially, the *Zytglogge* tower marked the western boundary of the city from 1191 until 1256, when the *Käfigturm* took over this role until 1345, which, in turn, was then succeeded by the *Christoffelturm* (located close to today's train station) until 1622. During the time of the Thirty Years' War two new fortifications, the so-called big and small *Schanze* (entrenchment), were built to protect the whole area of the peninsula.

Bern was occupied by French troops in 1798 during the French Revolutionary Wars, when it was stripped of parts of its territories. It regained the Bernese Oberland in 1802, and following the Congress of Vienna of 1814 newly acquired the Bernese Jura, once again becoming the largest canton of the confederacy as it stood during the Restoration, and further until the secession of the canton of Jura in 1979. In 1848 Bern was made the Federal City (seat of the Federal Assembly) of the new Swiss federal state.

A number of congresses of the socialist First and Second Internationals were held in Bern, particularly during World War I when Switzerland was neutral; see Berne International.

The city's population rose from about 5,000 in the 15th century to about 12,000 by 1800 and to above 60,000 by 1900, passing the 100,000 mark during the 1920s. Population peaked during the 1960s at 165,000, and has since decreased slightly, to below 130,000 by 2000. As of 31 December 2009, the resident population was at 130,289 of which 101,627 were Swiss citizens and 28,662 (22%) resident foreigners. Another estimated 350,000 people live in the immediate urban agglomeration.[10]

Geography

Bern lies on the Swiss plateau in the Canton of Bern, slightly west of the centre of Switzerland and 20 km (12 mi) north of the Bernese Alps. The countryside around Bern was formed by glaciers during the most recent Ice Age. The two mountains closest to Bern are the Gurten with a height of 958 m (3143 ft) and the Bantiger with a height of 947 m (3107 ft). The site of the old observatory in Bern is the point of origin of the CH1903 coordinate system at 46°57′08.66″N 7°26′22.50″E.

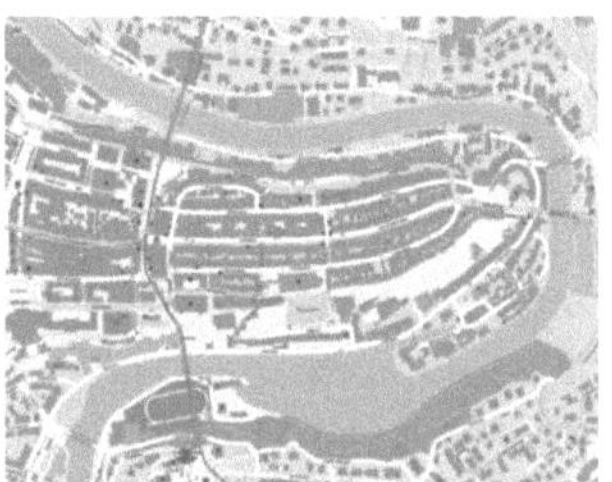
The Aar flows around the Old City of Bern with a loop east expansive

The city was originally built on a hilly peninsula surrounded by the river Aare, but outgrew the natural boundaries by the 19th century. A number of bridges have been built to allow the city to expand beyond the Aare.

Bern is built on very uneven ground. There are several dozen metres in difference of height between the inner city districts on the Aare (Matte, Marzili) and the higher ones (Kirchenfeld, Länggasse).

Bern has an area, as of 2009, of 51.62 square kilometers (19.93 sq mi). Of this area, 9.79 square kilometers (3.78 sq mi) or 19.0% is used for agricultural purposes, while 17.33 square kilometers (6.69 sq mi) or 33.6% is forested. Of the rest of the land, 23.25 square kilometers (8.98 sq mi) or 45.0% is settled (buildings or roads), 1.06 square kilometers (0.41 sq mi) or 2.1% is either rivers or lakes and 0.16 square kilometers (0.062 sq mi) or 0.3% is unproductive land.[11]

Of the built up area, 3.6% consists of industrial buildings, 21.7% housing and other buildings, and 12.6% is devoted to transportation infrastructure. Power and water infrastructure as well as other special developed areas made up 1.1% of the city, while another 6.0% consists of parks, green belts and sports fields. 32.8% of the total land area is heavily forested. Of the agricultural land, 14.3% is used for growing crops and 4.0% is designated to be used as pastures. The rivers and streams provide all the water in the municipality.[11]

Climate

Subdivisions

The municipality is administratively subdivided into six districts (*Stadtteile*), each of which consists of several quarters (*Quartiere*).

Demographics

Bern has a population (as of 31 December 2009) of 132'473, As of 2011, 24.7 % of the population was made up of foreign nationals. Most of the population (as of 2000) speaks German (81.2%), with Italian being second most common (3.9%) and French being third (3.6%).

52.7 % of the population are female, 47.3 % are male. The average age is 41 years and nine months. As of 2000 children and teenagers (0–19 years old) make up 15.1%, adults (20–64 years old) 65% and seniors (over 64 years old) 19.9%.

The Swiss population is generally well educated. In Bern about 72.8% of the population (between age 25–64) have completed either non-mandatory upper secondary education or additional higher education (either University or a *Fachhochschule*).

Bern has an unemployment rate of 3.2%. As of 2005, there were 773 people employed in the primary economic sector and about 104 businesses involved in this sector. 16,484 people are employed in the secondary sector and there are 1,094 businesses in this sector. 131,659 people are employed in the tertiary sector, with 7,638 businesses in this sector.[13]

Politics

Bern is governed by the *Gemeinderat*, an executive council with five members, one of them the elected mayor (*Stadtpräsident*). The parliament has 80 members and is called *Stadtrat*. Both the legislative and the executive are elected in general elections for a term of four years. The last elections were held in November 2008 with a 43.48% participation.

Erlacherhof

The executive council has a left-green majority with two representatives, including the mayor Alexander Tschäppät, of the *Social Democratic Party of Switzerland (SPS)* and one representative of the leftist Green party *Grünes Bündnis (GB)*. It also has a majority of three women against two men.

The seat of the *Gemeinderat* is the Erlacherhof.

Rathaus

The 80 members of the legislative council belong to 18 different political parties, the strongest being the *Social Democratic Party* with 20 representatives, followed by the conservative *Free Democratic Party of Switzerland (FDP)* with 10 and the moderate Green party *Grüne Freien Liste (GFL)* with 9 seats. Both the far right *Swiss People's Party (SVP)* and the leftist Green party *Grünes Bündnis* have 8 seats each.

The *Stadtrat* meets on Thursday evenings at the *Rathaus* (Town Hall).

The representatives of the *Social Democratic Party* and of the *Green Parties*, collectively referred to as "Red-Green-Center" (*Rot-Grüne-Mitte*), hold a majority in both councils and mostly determine City policy, although no formal coalition agreement exists and, under the system of direct democracy that prevails in Switzerland, most important issues are settled by general vote.

Main sights

The structure of Bern's city centre is largely medieval and has been recognised by UNESCO as a Cultural World Heritage Site. Perhaps its most famous sight is the *Zytglogge* (Bernese German for "Time Bell"), an elaborate medieval clock tower with moving puppets. It also has an impressive 15th century Gothic cathedral, the *Münster*, and a 15th century town hall. Thanks to 6 kilometres of arcades, the old town boasts one of the longest covered shopping promenades in Europe.

The Zytglogge clock tower and the city's medieval covered shopping promenades (*Lauben*)

Since the 16th century, the city has had a bear pit (the *Bärengraben*). The extended and renewed pit off the far end of the Nydeggbrücke actually contains four bears, including two young. During his visit in Bern in 2009, the Russian president and his wife gave two more young bears as a private present.[14] They are actually in Dählhölzli, Bern's zoo.

The Federal Palace (Bundeshaus), built from 1857 to 1902, which houses the national parliament, government and part of the federal administration, can also be visited.

Albert Einstein lived in an apartment at the Kramgasse 49, the site of the Einsteinhaus, from 1903 to 1905, the year in which the Annus Mirabilis Papers were published.

The Rose Garden (*Rosengarten*), from which a scenic panoramic view of the medieval town centre can be enjoyed, is a well-kept Rosarium on a hill, converted into a park from a former cemetery in 1913.

Bern's most recent sight is the set of fountains in front of the Federal Palace. It was inaugurated on August 1, 2004.

Bern features many heritage sites of national significance.[15] Apart from the entire Old Town and many sites within it, these include the Bärengraben, the Gewerbeschule Bern (1937), the Eidgenössisches Archiv für Denkmalpflege, the Kirchenfeld mansion district (after 1881), the Thunplatzbrunnen, the Federal Mint building, the Federal Archives, the Swiss National Library, the Historical Museum (1894), Alpine Museum, Museum of Communication and Natural History Museum.

The Universal Postal Union is situated in Bern.

Culture

Theatres

- Bern Theatre[16]
- Narrenpack Theater Bern[17]
- Schlachthaus-theater[18]
- Tojo Theater
- The Theater on the Effinger-Street[19]
- Theater am Käfigturm[20]

Zentrum Paul Klee

Cinemas

Bern has several dozen cinemas. As is customary in Switzerland, films are generally shown in their original language (e.g., English) with German and French subtitles. Only a small number of screenings are dubbed in German.

Film festivals

Stadttheater

- Queersicht – gay and lesbian film festival, held annually in the second week of November.
- SHNIT International Short Film Festival

Festivals

- BeJazz Summer and Winter Festival
- Buskers festival
- Gurtenfestival
- Internationales Jazzfestival Bern
- Queersicht – Queer Filmfestival, annually held second week of November.
- SHNIT International Short Film Festival
- Taktlos-Festival

Gurtenfestival, 2003

Fairs

- Zibelemärit – The Zibelemärit (onion market) is an annual fair held on the fourth Monday in November.

- Bernese Fassnacht (Carnival)

Sport

Bern was the site of the 1954 Football (Soccer) World Cup Final, a huge upset for the Hungarian Golden Team, who were beaten 3–2 by West Germany.

Stade de Suisse Wankdorf

The football team BSC Young Boys is based in Bern at the Stade de Suisse Wankdorf, which also was one of the venues for the European football championship 2008.

The *Stade de Suisse* hosted three matches during the 2008 UEFA Euro Cup tournament.

SC Bern is the major ice hockey team of Bern who plays at the PostFinance Arena.

The *PostFinance Arena* was the main host of the 2009 IIHF Ice Hockey World Championship, including the opening game and the final of the tournament.

The PostFinance Arena was also the host of the 2011 European Figure Skate Championships.

Bern Cardinals is the baseball and softball team of Bern, which plays at the Allmend

Bern Grizzlies is the American football club in Bern and plays at Sportanlage Schonau.

Bern was a candidate to host the 2010 Winter Olympics, but withdrew its bid in September 2002 after a referendum was passed that showed that the bid was not supported by locals. Those games were eventually awarded to Vancouver, Canada.

RC Bern is the local rugby club (since 1972) and plays at the Allmend. The ladies team has been founded in 1995.

Education

The University of Bern, whose buildings are mainly located in the *Länggasse* quarter, is located in Bern, as well as the University of Applied Sciences (*Fachhochschule*) and several vocations schools.

Transport

Bern is well connected to other cities by several motorways (A1, A12, A6).

Tram station on the Bahnhofplatz

Public transport works well in Bern, with tram, S-Bahn and bus lines which connect the different parts of the City. Bern Rail Station connects the City to the national and international train network. A funicular railway leads from the *Marzili* district to the *Bundeshaus*. This funicular is, with a length of 106 m (348 ft), the second shortest public railway in Europe after the Zagreb Funicular. Several Aare bridges connect the old parts of the city with the newer districts outside of the peninsula.

Bern is served by Bern Airport, located outside the city near the town of Belp. The regional airport, colloquially called *Bern-Belp* or *Belpmoos*, is connected to several Swiss and European cities.

Notable people

Albert Einstein's house

- Mikhail Bakunin died in Bern on 1 July 1876
- Albert Einstein worked out his theory of relativity while living in Bern, employed as a clerk at the patent office
- Albrecht von Haller
- Louise Elisabeth de Meuron, a famed eccentric and noble lady
- Paul Emmert, painter
- Ferdinand Hodler, painter
- Mark Streit, ice hockey player
- Christoph von Graffenried, founder of New Bern in the US state of North Carolina
- Peter Bieri, philosophy professor and novelist
- Adolf Wölfli, visual artist
- Roman Josi, ice hockey player
- Mani Matter, songwriter
- Léon Savary, Swiss writer and journalist
- Hans Urwyler, Christian minister

External links

- City of Bern [21]
- *Bern (Gemeinde)* in German [22], French [23] and Italian [24] in the online *Historical Dictionary of Switzerland.*
- GIS City of Bern [25]
- Bern Public Transportation Website (BernMobil) [26]
- Bern travel guide from Wikitravel
- CityHunter Bern [27]
- Gurtenfestival [28]

References

[1] Swiss Federal Statistical Office (http://www.bfs.admin.ch/bfs/portal/de/index/themen/01/02/blank/data/01.html), MS Excel document – *Bilanz der ständigen Wohnbevölkerung nach Kantonen, Bezirken und Gemeinden* (**German**) accessed 25 August 2010
[2] http://www.bern.ch
[3] http://www.bfs.admin.ch/bfs/portal/en/index/regionen/regionalportraets/gemeindesuche.html?geographyID=0351&FormEncoding=UTF-8
[4] http://www.bern.ch/leben_in_bern/stadt/statistik/in_kuerze
[5] Stadt Bern : *Medienmitteilung : 19 May 2003 : Eidgenössische Volkszählung 2000: Neue Definition der Agglomeration Bern (in [[German language|German* (http://www.bern.ch/leben_in_bern/stadt/statistik/volkszaehlung/bevoelkerung/MM18200311.pdf)])*]*
[6] Swiss Federal Statistical Office : *Population Size and Population Composition : Agglomerations : Permanent Resident Population in Urban and Rural Areas* (http://www.bfs.admin.ch/bfs/portal/en/index/themen/01/02/blank/key/raeumliche_verteilung/agglomerationen.html)
[7] http://www.are.admin.ch/themen/agglomeration/00641/03373/index.html?lang=fr
[8] of Living global city rankings – Mercer survey (http://www.mercer.com/referencecontent.htm?idContent=1173105#Top_50_cities:_Quality_of_living"Quality)
[9] Andres Kristol (ed.): Lexikon der schweizerischen Gemeindenamen. Huber, Frauenfeld 2005, ISBN 3-7193-1308-5, p. 143.
[10] municipal statistics, (http://www.bern.ch/leben_in_bern/stadt/statistik/in_kuerze) includes 6,816 weekend commuters not included in the federal statistics of 123,466. (http://www.bfs.admin.ch/bfs/portal/de/index/themen/01/02/blank/dos/result.html)
[11] Swiss Federal Statistical Office-Land Use Statistics (http://www.bfs.admin.ch/bfs/portal/de/index/themen/02/03/blank/data/gemeindedaten.html) 2009 data (**German**) accessed 25 March 2010
[12] "Average Values-Table, 1961–1990" (http://www.meteoswiss.admin.ch/web/de/klima/klima_schweiz/tabellen.html) (in German, French, Italian). Federal Office of Meteorology and Climatology MeteoSwiss. . Retrieved 8 May 2009.
[13] Swiss Federal Statistical Office (http://www.bfs.admin.ch/bfs/portal/en/index/regionen/regionalportraets/gemeindesuche.html) accessed 29-May-2009
[14] "City of bears receives Russian bruins" (http://www.swissinfo.ch/eng/swiss_news/City_of_bears_receives_Russian_bruins.html?cid=1008832). *swissinfo.ch*. 16 September 2009. .
[15] Swiss inventory of cultural property of national and regional significance (1995), p. 103–105.
[16] "Stadttheater Bern" (http://www.stadttheaterbern.ch/). . Retrieved 2009-04-12.
[17] "Narrenpack Theatre Bern" (http://www.narrenpack.ch/). . Retrieved 2009-04-12.
[18] "Schlachthaus Theatre Bern" (http://www.schlachthaus.ch/). . Retrieved 2009-04-12.
[19] "Das Theatre an der Effingerstrasse" (http://www.dastheater-effingerstr.ch/). . Retrieved 2009-04-12.
[20] "Theater am Käfigturm" (http://www.theater-am-kaefigturm.ch/). . Retrieved 2009-04-12.
[21] http://www.bern.ch/
[22] http://www.hls-dhs-dss.ch/textes/d/D209.php
[23] http://www.hls-dhs-dss.ch/textes/f/F209.php
[24] http://www.hls-dhs-dss.ch/textes/i/I209.php
[25] http://web.archive.org/web/20071023192334/http://www.geobern.ch/TBInternet/default.aspx?Show=Bern&Lang=en
[26] http://www.bernmobil.ch/
[27] http://www.cityhunter.ch/
[28] http://www.gurtenfestival.ch/

Grindelwald Grund railway station

Grindelwald Grund regional rail	
Grindelwald Grund railway station	
Station statistics	
Coordinates	46°37′22″N 08°01′24″E
Lines	Wengernalpbahn
Platforms	2
Other information	
Opened	1893
Electrified	24 June 1910

Grindelwald Grund railway station is a station on the Wengernalpbahn in the Swiss canton of Bern.

Next station South: **Brandegg**	**Wengernalpbahn**	Next station North: **Grindelwald**

Schwendi railway station

Schwendi regional rail	
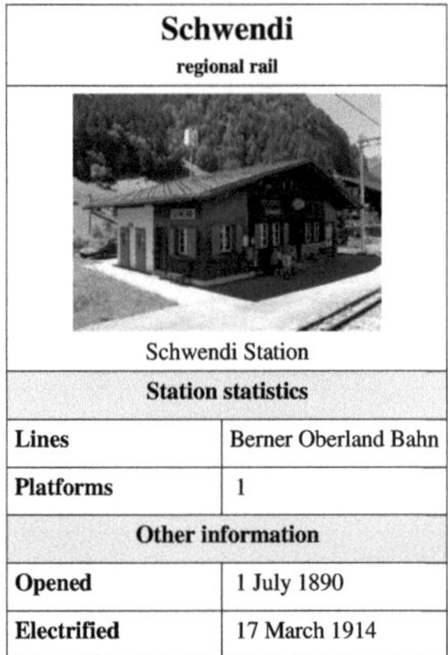 Schwendi Station	
Station statistics	
Lines	Berner Oberland Bahn
Platforms	1
Other information	
Opened	1 July 1890
Electrified	17 March 1914

Schwendi railway station serves the village of Schwendi in the Swiss canton of Bern.

Next station East: **Grindelwald**	Berner Oberland Bahn	Next station West: **Burglaunen**

Wengernalpbahn

Wengernalpbahn	
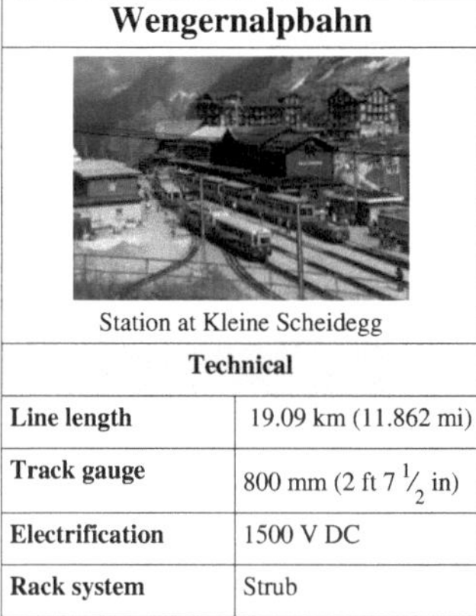 Station at Kleine Scheidegg	
Technical	
Line length	19.09 km (11.862 mi)
Track gauge	800 mm (2 ft 7 $^1/_2$ in)
Electrification	1500 V DC
Rack system	Strub

Wengernalpbahn (or **WAB**) is a 19.091 km long, 800 mm (2 ft 7 $^1/_2$ in) gauge rack railway line in Switzerland, which runs from Lauterbrunnen and Grindelwald (two branches) to Kleine Scheidegg, making it the world's longest continuous rack and pinion railway.

At Kleine Scheidegg passengers transfer to the Jungfraubahn for the continuation of the journey to the highest railway station in Europe at Jungfraujoch.

Timeline

Diamond crossing at Grindelwald Grund. The Strub rack is shown clearly, with a short Riggenbach section across the crossing.

- 1875 First plans for the Berner Bahngesellschaft were drawn up but the high projected costs meant that the concession expired.
- 1890 Leo Heer-Bétrix gained a new 80 year concession to build and operate the railway. The Wengernalp Bahn was founded.
- 1891 Construction work started.
- 1892 The first steam locomotive reaches Wengen on 18 April and Kleine Scheidegg on 10 August.
- 1893 The first line from Lauterbrunnen to Grindelwald opened on 20 June as a summer only service. The racks are of the Riggenbach rack system, as modified by Arnold Pauli. Modern stock uses Strub rack system.
- 1909 Originally operated by steam locomotives, electrification of the line from Lauterbrunnen to Kleine Scheidegg was completed on 3 June using 1500 V DC, with electric locomotives positioned, for safety reasons, at the lower end of the trains.
- 1910 Electrification of section between Grindelwald and Kleine Scheidegg completed on 24 June. Opening of the longer but less steep Lauterbrunnen to Wengen route on 7 July.
- 1912 Steam operation ceased.

- 1913 Winter operations from Lauterbrunnen to Kleine Scheidegg started.
- 1925 Year round from Lauterbrunnen to Kleine Scheidegg started.
- 1934 Winter operations from Grindelwald to Kleine Scheidegg started.
- 1942 The headquarters of the railway moves from Zurich to Interlaken.
- 1947 The first 3 motorcoaches are purchased.

Triangular junction at Kleine Scheidegg, partly built into the mountain.

- 1948 The turning triangle at Kleine Scheidegg is constructed.
- 1960 Year round operations from Grindelwald to Kleine Scheidegg started.
- 1984 First class services were removed and since then the railway has only offered one class of service.
- 1990 An avalanche shelter is built on the Lauterbrunnen side of the operation.
- 1995 Wengen station is rebuilt to include a freight delivery terminal.
- 2005 Wengernalp station platforms were extended from 127 m to 181 m usable length. The station extension went into operation on 16 November. Total costs were around CHF 950,000.
- 2007 Plans are announced to decommission and dismantle the old line between Lauterbrunnen and Wengen because of the steep gradient of 25% and the geological security risks.
- 2009 The track is lifted on the old line between Lauterbrunnen and Wengen.

Operations

Nowadays, most passenger trains are made up of railcars, the powered car still being positioned at the lower end of the train, and so train compositions do not usually cross Kleine Scheidegg to travel directly from Lauterbrunnen over to Grindelwald. However, a triangular junction specially built into the mountainside at Kleine Scheidegg allows the train to be turned if necessary so that it can also be used on the other side of the col. The newest of these trains reach 28 km/h on the steepest stretch.

At peak periods, additional trains can be put into operation at short intervals ahead of the scheduled train, allowing capacity to be optimised according to demand. This demands an extremely flexible organization procedure and enormous care and attention with regard to dispatching trains.

The busiest stretch of railway runs from Lauterbrunnen to Wengen as this is also used to transport goods to traffic-free Wengen.

The railway operates two workshops at Lauterbrunnen and Grindelwald Grund.

Stations

- Lauterbrunnen - connection to the Berner Oberland Bahn for Interlaken Ost and the Bergbahn Lauterbrunnen-Mürren to Mürren. Also the home of one of the two depots of the Wengernalpbahn.
- Wengwald - request stop
- Wengen - connection to the Männlichen cable car Luftseilbahn Wengen-Männlichen. With 4 through platforms, a bay platform facing Lauterbrunnen and a freight depot underneath the passenger station, this is the largest rack railway station in the world.
- Allmend - request stop
- Wengernalp

- Kleine Scheidegg - connection to the Jungfraubahn and terminus of services from Lauterbrunnen and Grindelwald.
- Alpiglen - request stop
- Brandegg - request stop
- Grindelwald Grund - connection to the Gondelbahn Grindelwald-Männlichen. Trains travelling between Grindelwald and Kleine Scheidegg reverse at Grindelwald Grund. Also the home of one of the two depots of the Wengernalpbahn. Train announcements and some station signs are just labelled **Grund** to avoid confusion with Grindelwald.
- Grindelwald - connection to the Berner Oberland Bahn for Interlaken Ost.

Rolling stock

Locomotive 32 pushes a freight train approaching Wengen with the old original line just visible approaching the new one.

Locomotives

No.	Class	Builders	Date Completed
31	He2/2	Stadler/SLM/BBC	1995
32	He2/2	Stadler/SLM/BBC	1995
51	He2/2	SLM/Alioth	1909
52	He2/2	SLM/Alioth	1909
53	He2/2	SLM/Alioth	1909
54	He2/2	SLM/Alioth	1909
64	He2/2	SLM/BBC	1926
65	He2/2	SLM/MFO	1929

Railcars

The earliest railcars still in service, numbers 101 and 102 (BDhe4/4, built by SLM/BBC) date from 1947, No.102 being refurbished 1985. Of the other 14 cars of this class, No's. 104 and 106 - 118, built between 1954 and 1964, which still survive, ten have been refurbished (all except 104 / 7 / 9 / 10) since 2000. No.114 was refurbished along with 102 in 1985 but was returned to works in 2002 to be brought up to the newer specifications.

In 1970 a new class of railcar (BDhe4/4) was introduced, numbered 119 to 124, built by a consortium of SIG, SLM, SAAS [1] and BBC. These were refurbished in 1998. Four further cars, class BDhe4/8, numbered 131 to 134, and built by SLM (Works Nos. 5363 - 66 inclusive) with electrical equipment by BBC, arrived in 1988 with the latest additions, a series of four "Panorama" cars of Class Bhe4/8, built by Stadler arriving in 2004.

Driving trailer cars

The earliest coaches still listed to the company are three dating from 1893 and a single example from 1901. These have all been rebuilt twice, the final one in 1995 by the von Roll company. Later examples date from the period 1959 to 2003, some of the earlier of these have rebuilt. Building of the stock has been carried out by SIG, with electrical equipment by Brown Boveri / Asea Brown Boveri and later by Stadler with electrical equipment by Steck, (except No. 231 which was built by SLM with electrical equipment by ABB).

The line also operates with an extensive collection of goods stock, in the main used for services between Lauterbrunnen and Wengen.

See also

- Rail transport in Switzerland
- Jungfraubahn

External links

- Jungfrau Railways website (English) [2]

References

[1] http://www.ub.unibas.ch/wwz/vsa/106-008.htm
[2] http://www.jungfraubahn.ch/en/DesktopDefault.aspx/tabid-1/

Berner Oberland Bahn

Berner Oberland Bahn	
 A BOB train headed by class ABeh4/4 No.312 (named *Interlaken*) at Interlaken Ost, note the former brown / cream livery.	
Technical	
No. of tracks	mostly single track with passing points and a double track section at the lower end.
Track gauge	1000 mm (3 ft 3 $\frac{3}{8}$ in)
Maximum incline	12 %
Rack system	Riggenbach

The **Berner Oberland Bahn** (BOB, pro. beh-oh-beh) is a narrow-gauge mountain railway in the Bernese Oberland region of Switzerland. It runs, via a "Y" junction at Zweilütschinen to serve Interlaken and Lauterbrunnen and Grindelwald. The railway is rack assisted (that is although an adhesion railway, rack and pinion operation is used on steep sections of the line to assist traction). Also part of BOB is the 800 mm (2 ft 7 $\frac{1}{2}$ in) Schynige Platte Railway.

History

Planning

The first proposals for the Berner Oberland Bahn, made in 1873, showed a line from Interlaken (at that time Aarmühle) to Zweilütschinen with later options to Lauterbrunnen and Grindlewald with starting point at Bönigen. Four years later an 80 years concession was obtained for construction and operation of the line and the company, Berner Oberland-Bahn was founded on 2 November 1888 and construction started the following year

Failure of the plan to extend to Visp

In 1897 the company obtained a concession to construct a 54.7 km line from Lauterbrunnen to Visp, with stations at Stechelberg, Steinberg, Oberborn, and Blattern. It would have involved the construction of a 4,650 m tunnel at 2,200 m elevation under the Breithorn mountain. At Visp it would have had a connection with the Simplon line.

Estimated at 15 million Swiss francs, finance was not forthcoming and by 1906 the plans were abandoned.

Initial operations

By 1 July 1890 the 1000 mm (3 ft 3 $^3/_8$ in) gauge line, was opened, using steam traction.

On 18 August 1902 a disastrous fire destroyed the station buildings and goods shed at Grindelwald and these were later rebuilt, surviving to the present day. On 7 October 1908 a new station was added to the system, that at Schwendi on the Grindelwald section.

Steam traction on the line came to an end in 1914, the line becoming electrified at 1500 V d.c., overhead supply, on 17 March of that year, although steam locomotives have been used since that date on special services.

Several changes were made during the 1950s and 1960s, the two most important being in 1957, the construction of an airfield at Interlaken causing the realignment of the line between Wilderswil and Interlaken Ost, but to no detriment and, with a need for servicing and construction facilities on the line a new depot was opened at Zweilütschinen in 1968.

Recent improvements

Since that time there has been a need for track capacity to be increased and in 1991 the Wilderswil to Gsteigwiler section was substantially improved. This was followed by the doubling of the Gsteigwiler to Zweilütschinen 4 years later. A bottleneck between Wilderswil and Zweilütschinen was eased when, in 1999, a 2.5 km. double track section was opened between those places meaning that trains could run through without the need to use the passing loop and, as necessary, awaiting the train in the opposite direction.

A modern low-floor train of the BOB in Grindelwald with the track of the Wengernalpbahn on the adjacent platform. Note the modern blue / yellow BOB livery.

The BOB has a total length of 23.608 km and is a mixed rack and adhesion railway with four rack and pinion sections, using the Riggenbach rack system system, two each on the steep sections of both arms of the line.

Fatal accident in 2003

On 7 August 2003 two trains collided head-on on a single track section between Zweilütschinen and Wilderswil, 1 person was killed and 64 injured.[1] The regular train coming down from Zweilütschinen had passed a red signal at the end of the double track section and collided with an extra train near Gsteigwiler. Automatic train stop system ZSI-127 had already been in place but not yet in use, awaiting final completion and approval.

Operations

Since 1949 railcars have predominated. Some of the older electric locomotives still survive and are used for special trains. The centre of operations is Zweilütschinen with the depot headquarters and the modern main workshops.

A train at the Wilderswil station with the track of the Schynige Platte Railway (red train) on the adjacent platform

From the introduction of the 1999 timetable, the newly constructed 2.5 km section of dual track between Gsteigwiler and Zweilütschinen allows trains to pass without one having to wait in a loop, off the main line. This means that a half-hour timetable can be operated with only five train compositions. Since 2005, every composition has been equipped with an articulated (three-part) low-floor driving trailer as standard.

Two train compositions are usually coupled together to travel to Zweilütschinen where they are then split. The front portion travels to Lauterbrunnen, the other one to Grindelwald. The motor coach (power unit) is always positioned on the uphill side, a driving trailer (coach with a driver's cab) being positioned on the downhill side, to avoid any running round manoeuvres at the terminus stations.

Stations

Station	Distance (km)	Height (m)	Information
Interlaken Ost	-0.18	567	connections to the Zentralbahn, BLS and Swiss Federal Railways
Wilderswil	3.24	584	connections to the Schynige Platte Railway
Zweilütschinen	8.18	652	trains divide with front portion for Lauterbrunnen and the rear portion for Grindelwald
Sandweid	10.42	725	Request stop
Lauterbrunnen	12.28	795	connections to the Wengernalpbahn for Kleine Scheidegg via Wengen and the Bergbahn Lauterbrunnen-Mürren to Mürren
Lutschental	12.29	714	Request stop
Burglauenen	14.43	896	Request stop
Schwendi	16.82	920	Request stop
Grindelwald	19.41	1034	connections to the Wengernalpbahn for Kleine Scheidegg

Timetable

Like all Swiss railways the BOB operates to a clock - face timetable offering connections from the main line at Interlaken and, at its upper terminals, to the Wengeralpbahn.

Throughout the year the BOB offers a service every hour in each direction on both its lines, the trains leaving Interlaken Ost coupled together and dividing at Zweilütschinen. These are compilmented in the summer months additional trains to give a 30 minutes service in the morning and afternoon. The last services are often timetabled to be operated by buses.

The departure board at Sandweid station

Locomotives / Railcars

No.	Name	Class	Seats: 1st/2nd	Builders Details	Date Built	Notes.
1		Tm		Stadler/Sr/BBC/ MFO/SIG	1946	Rebuilt 1980; 6-cyl/110 kW
21		Xm1/2		P&T	1979	6-cyl diesel/123 kW
24		HGe3/3		SLM/MFO/BBC	1914	Rebuilt 1940
29		HGe3/3		SLM/MFO	1926	
31		HGm2/2		Steck/Deutz/SLM	1985	6-cyl diesel/296 kW
301		ABDeh4/4	10/32	SLM/BBC	1949	Leased to MIB, 1995 scrapped
302		ABDeh4/4	10/32	SLM/BBC	1949	Engineers Dept.
303		ABDeh4/4	10/32	SLM/BBC	1949	Engineers Dept.
304		ABeh4/4	12/32	SIG/SLM/BBC	1965	Brown/Cream livery at 9.2007.
305	*Gündlischwand*	ABeh4/4	12/32	SIG/SLM/BBC	1965	Rebuilt 1998
306	*Lütschental*	ABeh4/4	12/32	SIG/SLM/BBC	1965	Rebuilt 1997
307	*Wilderswil*	ABeh4/4	12/32	SIG/SLM/BBC	1965	Rebuilt 2002
308	*Gsteigwiler*	ABeh4/4	12/32	SIG/SLM/BBC	1979	
309		ABeh4/4	12/32	SIG/SLM/BBC	1979	1999 sold to BZB
310	*Matten*	ABeh4/4	12/32	SIG/SLM/BBC	1979	Rebuilt 2007
311	*Grindelwald*	ABeh4/4	12/24	SLM 5296/BBC	1986	
312	*Interlaken*	ABeh4/4	12/24	SLM 5297/BBC	1986	
313	*Lauterbrunnen*	ABeh4/4	12/24	SLM 5298/BBC	1986	
(321)		BDe4/4	0/34	SIG/SAAS	1953	2003 ex-CJ No.601*), 2006 sold to LEB No.28**)
(322)		BDe4/4	0/34	SIG/SAAS	1953	2003 ex-CJ No.604*), 2005 sold to MIB No.10

- Rebuild 1997-2008 of 304-310 included fitting of equipment for push-pull-trains, available from delivery on 311-313
- *) for use on the then-planned but finally not built branch line to Mystery Park
- **) arrived on LEB still numbered 601 on 22 February 2006 and was used together with Bt 702 arriving directly from CJ. LEB finally purchased the two vehicles.

Rolling stock

The passenger rolling stock of the line can be divided into that in regular use and that which is historic in nature. Present day passenger stock is painted in striking a blue/yellow livery.

A train in Lauterbrunnen with Stadler ABt, low floor 3-car set nearest the camers.

That in regular use can be divided as follows:

- Series A, First class open saloon bogie coaches with 36 seats, numbered 181 and 182, built by SIG in 1971 ,with 182 being rebuilt in 1999, MU-wired, and 181 sold to SBB, now Zentralbahn A 217.
- Series AB, First/Second Composite bogie saloon open platform coaches with 18 first and 48 second class seats, originally numbered 205-210, built by SIG and delivered, the first two in 1952, the second pair in 1954 and the final pair in 1956. No.209 is the only member to be found on the line and still carrying (Oct 2009) the Brown and Cream BOB livery. It is not regularly working and can usually be found in Interlaken. (See Preservation Notes (Below).
- Series AB, First/Second Composite bogie saloon coaches with 24 first and 23 second class seats, numbered 211-215, built by SIG and delivered in 1970. All have since been fitted with MU wiring, 211, 213 and 214 sold to Zentralbahn.
- Series AB, First/Second Composite bogie coach with 23 first and 22 second class seats, numbered 221, built by SIG in 1946 and purchased from SBB (Brünig) in 1997, scrapped 2006.
- Series B, Second class open saloon bogie coaches with 72 seats, numbered 232-237, built by SIG between 1952 and 1956. Five members of this group are preserved,(See notes below). Number 232 is still on the line and shown in stock lists but is presently (2009) to be found in Interlaken in Brown/Cream livery.
- Series B, Second class saloon open platform bogie coaches with 64 seats, numbered 241-256, built by SIG between 1967 and 1970, most have been rebuilt but five members have been scrapped. 253-256 originally Zentralbahn. Repainted and MU-wired: 241, 245, 247, 250-256. 242 was MU-wired but is still in Brown/Cream livery and permantly sitting in Interlaken.
- Series B, Second class open saloon bogie coaches with 62 seats, numbered 261 and 262, built by ACMV/SIG and delivered in 1987. MU-wired.
- Series B, Second class open saloon, bogie coach with 52 seats, numbered 271, built by SIG for the SBB in 1954 as AB477 and rebuilt in 2001, scrapped 2006.
- Series B, Second class open saloon bogie coaches with 60 seats, numbered 271-274, built by SIG for the SBB as B861/B863/B846 and delivered in 1954, being purchased and repainted in 1998, scrapped 1999/2006/2003/2006.
- Series BD, Second class open saloon bogie coaches (40 seats) with guards/parcels compartment, numbered 501-503. These were built by SIG and delivered to the SBB, adapted for push-pull trains as numbers 512-4 / 511-6 / 510-8 in 1968/9. They were rebuilt 2003/4.
- Series BDt, Second class driving trailer (40 seats) with guards/parcel compartment, numbered 401-403, built by ACMV/SIG/BBC in 1987.
- Series ABt, First/Second composite driving trailer with 18 first and 31 second class seats, numbered 411 to 415 inclusive, built for the RBS as ABt 207/3/6/4/5 by FFA/SWP in 1982, being rebuilt by the BOB in 2003–06.

- Series ABt, First/Second composite driving trailer three-car sets with low floor access built by Stadler. These were delivered in 2005 and now form part of every train. They are numbered from 421 to 425 inclusive.
- Series D, Guards/Luggage bogie coach built by SIG in 1971 numbered 531 to 535 inclusive. 532 sold to CJ, 535 scrapped, 532 MU wired, still brown/cream

Historical stock includes the following items, which still carry the former brown/cream livery for coaches and all-over brown for guards/parcels vehicles.

- Series A3, First Class saloon, double verandha with 30 seats, No.102. repairs required to one end due to accident.
- Series BC4, First/second class open saloon coach with 14 first and 38 second class seats, numbered 203, built by SIG in 1938 and rebuilt in 1988.
- Series C3, Second class open saloon coach numbered 29.
- Series D3, Guards/parcels carriage, numbered 515 and 516, built by SIG in 1911.

Goods stock is a varied collection, much of which would not be out of place in a museum. The earliest wagon shown on the BOB stock list dates from 1888 and was rebuilt by the BOB in 1990. The collection of goods stock totals over 30 assorted wagons, most pre-First World War, many built by SIG and much rebuilt by the BOB over the years. More recently a few additions have been made, most of which are second-hand from CFF/SBB/FFS. The line is home to a snowplough (Series Xrot e) with was built in 1954 by SIG/BBC and rebuilt in 1990 at the BOB workshops.

Preservation

Several items of rolling stock have been sold (transferred) to metre gauge preserved railways.

- Series B, Second class open saloon bogie coaches with 52 seats, numbered 201 and 202, built by SIG in 1930 are preserved by the La Traction group. Both were rebuilt, 201 in 1965, 202 in 1972 and both again in 1997.
- Series AB, First/Second Composite saloon bogie coach No. 204, with 24 first and 23 second class seats, built by SIG in 1938 and rebuilt in 1997 can also be found at the depot of La Traction.
- Series AB, First/Second Composite saloon bogiecoaches with 18 first and 48 second class seats, numbered 205-210, built by SIG and delivered, the first two in 1952, the second pair in 1954 and the final pair in 1956. All, except 206 and 208 have been rebuilt. No.205 is preserved and works on the Brohltalbahn whilst 207,208 and 210 can be found on the Chemin de Fer de la Baie de Somme (CFBS).
- Series B has five preserved members, No's 234 and 235 are to be found on the Brohltalbahn with 205, whilst 231, 236 and 237 are in northern France working on the CFBS.
- Series D, Guards / parcels carriages, No.521, built by SIG in 1916 and rebuilt by BOB in 1973/4 can be found working on the Brohltalbahn, whilst No 522, again built by SIG in 1916 and rebuilt by BOB in 1973/4 together with No.523, built in 1908 by SIG and rebuilt by BOB in 1976 are to be found on the CFBS. Vehicle No. 522 is undergoing, (at Spring 2007), a rebuilt into a catering car for use on the CFBS dining car train.

References

[1] Swiss Info (http://www.swissinfo.org/sen/swissinfo.html?siteSect=105&sid=4105186)

Sources

Items shown in the above list are taken from official BOB listings, last issue September 2004, and have been updated by personal observations made during September 2007.

External links

- BOB on Rail info (http://www.rail-info.ch/BOB/index.en.html)

Schynige Platte

Schynige Platte	
Schynige Platte view (the railway station is visible)	
Elevation	2076 m (6811 ft)
Location	
Schynige Platte	
Range	Bernese Alps
Coordinates	46°39′22″N 7°54′25″E
Climbing	
Easiest route	Railway

The **Schynige Platte** is a Swiss mountain region near Wilderswil where the Schynige Platte Railway ends (at 1970 meters above sea level).In good weather conditions the summit offers spectacular views to many surrounding mountains, including Jungfrau, Eiger, Silberhorn and others, as well as Thunersee and Brienzersee. An Alpine Garden is located near the summit where various alpine plants can be found together with labels, indicating their Latin names (many of these plants also grow around as wild species). At the top of the railway station there is a hotel and mountain restaurant.

Schynige Platte is a start point for the popular hiking trails to Faulhorn or First.

External links

- Up to Schynige Platte [1]

References

[1] http://addiator.blogspot.com/2009/08/up-to-schynige-platte.html

Wilderswil

Wilderswil	

Country Switzerland **Canton** Bern **District** Interlaken-Oberhasli 46°40′N 7°52′E	
Population	2436 (Dec 2009)[1]
- Density	180 /km^2 (467 /sq mi)
Area	13.5 km^2 (5.2 sq mi)
Elevation	586 m (1923 ft)
Postal code	3812
SFOS number	0594
Mayor	M Eduard Schild
Surrounded by	Bönigen, Därligen, Gsteigwiler, Gündlischwand, Lauterbrunnen, Matten bei Interlaken, Saxeten
Website	www.wilderswil.ch [2] SFSO statistics [3]
Wilderswil	
View map of Wilderswil	

Wilderswil is a municipality in the Interlaken-Oberhasli administrative district in the canton of Bern in Switzerland.

It is situated at the southern border of the Bödeli watershed, where the Saxetenbach joins the Lütschine, 4 km (2.5 mi) south of Interlaken. The village is situated at the foot of the mountains Eiger, Mönch, and Jungfrau. The highest point in the municipality is 2413 m (7917 ft).

History

The area was settled by Alamanni around the year 600. The name also comes from this period. In 1895, in excavations for the construction of a hotel, a graveyard of 15 graves, with 18 skeletons and burial objects, was discovered.

Wilderswil is first mentioned in the year 1224. The place was governed by the barons of Rotenfluh-Wilderswil, later also by the barons of Waediswil, Weissenburg and Scharnachtal. In about 1500 the city of Bern took over.

Tourism

Wilderwil

Because of its location it is the point of departure for many excursions to the heart of the Jungfrau region. The town has 16 hotels, motels and hotels with 900 guest beds, 300 vacation homes, and one camping site (open in the summer).

The town is easily accessed by the A8 motorway or by frequent trains from Interlaken.

Wilderswil railway station is the terminus of the Schynige Platte Railway and also a station on the Berner Oberland Bahn.

Geography

Wilderswil has an area of 13.5 km^2 (5.2 sq mi). Of this area, 21.7% is used for agricultural purposes, while 55.9% is forested. Of the rest of the land, 7.9% is settled (buildings or roads) and the remainder (14.5%) is non-productive (rivers, glaciers or mountains).[4]

Demographics

Wilderswil has a population (as of 31 December 2009) of 2436.[1] As of 2007, 8.3% of the population was made up of foreign nationals. Over the last 10 years the population has grown at a rate of 18.5%. Most of the population (as of 2000) speaks German (94.5%), with Serbo-Croatian being second most common (1.0%) and Italian being third (0.9%).

In the 2007 election the most popular party was the SVP which received 35.4% of the vote. The next three most popular parties were the FDP (18.4%), the SPS (16.7%) and the Green Party (12.3%).

The age distribution of the population (as of 2000) is children and teenagers (0–19 years old) make up 25% of the population, while adults (20–64 years old) make up 59.6% and seniors (over 64 years old) make up 15.5%. The entire Swiss population is generally well educated. In Wilderswil about 78.7% of the population (between age 25-64) have completed either non-mandatory upper secondary education or additional higher education (either University or a *Fachhochschule*).

Wilderswil has an unemployment rate of 2.69%. As of 2005, there were 43 people employed in the primary economic sector and about 21 businesses involved in this sector. 375 people are employed in the secondary sector and there are 24 businesses in this sector. 484 people are employed in the tertiary sector, with 75 businesses in this sector.[4]

Famous people

- Urs Räber, skier

References

[1] Swiss Federal Statistical Office (http://www.bfs.admin.ch/bfs/portal/de/index/themen/01/02/blank/data/01.html), MS Excel document – *Bilanz der ständigen Wohnbevölkerung nach Kantonen, Bezirken und Gemeinden* (German) accessed 25 August 2010

[2] http://www.wilderswil.ch

[3] http://www.bfs.admin.ch/bfs/portal/en/index/regionen/regionalportraets/gemeindesuche.html?geographyID=0594&FormEncoding=UTF-8

[4] Swiss Federal Statistical Office (http://www.bfs.admin.ch/bfs/portal/en/index/regionen/regionalportraets/gemeindesuche.html) accessed 01-Jul-2009

Wilderswil railway station

Wilderswil regional rail	
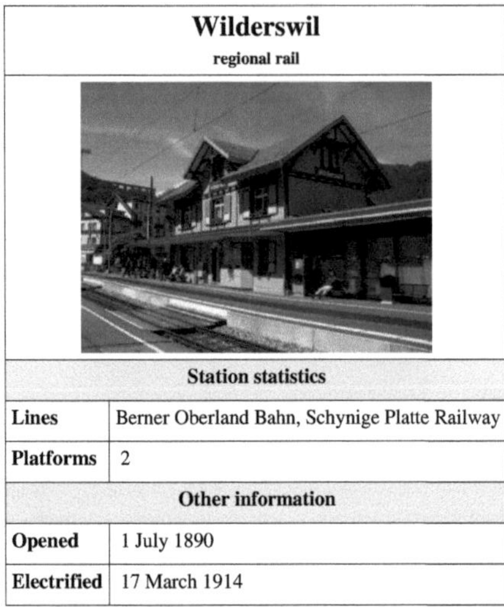	
Station statistics	
Lines	Berner Oberland Bahn, Schynige Platte Railway
Platforms	2
Other information	
Opened	1 July 1890
Electrified	17 March 1914

Wilderswil railway station serves the town of Wilderswil in the Swiss canton of Bern.

Berner Oberland Bahn trains operate services to Interlaken Ost, Grindelwald and Lauterbrunnen. It is also the valley terminus for the Schynige Platte Railway, whose trains are stabled at a depot bordering the station. There are also bus services to Interlaken West via Matten bei Interlaken.

Next station South: **Zweilütschinen**	Berner Oberland Bahn	Next station North: **Interlaken Ost**

Lauterbrunnen

Lauterbrunnen	

Country	Switzerland
Canton	Bern
District	Interlaken-Oberhasli
46°36′N 7°54′E	
Population	2504 (Dec 2009)[1]
- Density	15 /km² (39 /sq mi)
Area	164.4 km² (63.5 sq mi)
Elevation	795 m (2608 ft)
Postal code	3822
SFOS number	0584
Localities	Lauterbrunnen, Wengen, Mürren, Gimmelwald, Stechelberg, Isenfluh
Surrounded by	Aeschi bei Spiez, Blatten (Lötschen) (VS), Fieschertal (VS), Grindelwald, Gündlischwand, Kandersteg, Lütschental, Reichenbach im Kandertal, Saxeten, Wilderswil
Website	www.lauterbrunnen.ch [2] SFSO statistics [3]

Lauterbrunnen	

View map of Lauterbrunnen

Lauterbrunnen is a municipality in the Interlaken-Oberhasli administrative district in the canton of Bern in Switzerland.

The municipality lies in the Lauterbrunnen Valley and comprises the villages Lauterbrunnen, Wengen, Mürren, Gimmelwald, Stechelberg and Isenfluh. The population of the Lauterbrunnen village is less than that of Wengen, but greater than that of the others.

History

Lauterbrunnen is first mentioned in 1240 as *in claro fonte*. In 1304 it was mentioned as *Luterbrunnen*.[4]

Geography

View of the valley from the Männlichen

Lauterbrunnen has an area, as of 2009, of 164.56 square kilometers (63.54 sq mi). Of this area, 36.79 square kilometers (14.20 sq mi) or 22.4% is used for agricultural purposes, while 28.84 square kilometers (11.14 sq mi) or 17.5% is forested. Of the rest of the land, 2.31 square kilometers (0.89 sq mi) or 1.4% is settled (buildings or roads), 1.08 square kilometers (0.42 sq mi) or 0.7% is either rivers or lakes and 95.39 square kilometers (36.83 sq mi) or 58.0% is unproductive land.[5]

Of the built up area, housing and buildings made up 0.7% and transportation infrastructure made up 0.5%. 13.6% of the total land area is heavily forested and 2.0% is covered with orchards or small clusters of trees. Of the agricultural land, 3.5% is pastures and 18.9% is used for alpine pastures. All the water in the municipality is in rivers and streams. Of the unproductive areas, 10.3% is unproductive vegetation, 31.3% is too rocky for vegetation and 16.3% of the land is covered by glaciers.[5]

The river *Weisse Lütschine* flows through Lauterbrunnen and overflows its banks about once a year. The source of the river comes from melting snow high in the mountains, thus making it a very pure and clean source of water. It is common practice in the camp sites to chill drinks in the water.

Lauterbrunnen lies at the bottom of a hanging or U-shaped valley that extends south and then south-westwards from the village to meet the 8 kilometers (5.0 mi) Lauterbrunnen Wall. The valley of Lauterbrunnen (*Lauterbrunnental*) is one of the deepest in the Alpine chain when compared with the height of the mountains that rise directly on either side. It is a true cleft, rarely more than one kilometre in width, between limestones precipices, sometimes quite perpendicular, everywhere of extreme steepness. It is to this form of the valley that it owes the numerous waterfalls from which it derives its name. The streams descending from the adjoining mountains, on reaching the verge of the rocky walls of the valley, form cascades so high that they are almost lost in spray before they reach the level of the valley. The most famous of these are the Staubbach Falls within less than one kilometres of the village of Lauterbrunnen. The height of the cascade is between 800 and 900 feet (240 and 270 m), one of the highest in Europe formed of a single unbroken fall.[6]

Demographics

The Lauterbrunnen Valley. The village of Lauterbrunnen (foreground), the Staubbach Falls (centre right), the Jungfrau (top left) and the Lauterbrunnen Wall (background).

Lauterbrunnen has a population (as of 31 December 2009) of 2504.[1] As of 2007, 19.2% of the population was made up of foreign nationals. Over the last 10 years the population has decreased at a rate of -14.3%. Most of the population (as of 2000) speaks German (85.2%), with Portuguese being second most common (4.9%) and Serbo-Croatian being third (2.0%).

In the 2007 election the most popular party was the SVP which received 38.1% of the vote. The next three most popular parties were the FDP (19.8%), the SPS (14.5%) and the Green Party (10.5%).

The age distribution of the population (as of 2000) is children and teenagers (0–19 years old) make up 22.4% of the population, while adults (20–64 years old) make up 58.9% and seniors (over 64 years old) make up 18.7%. The entire Swiss population is generally well educated. In Lauterbrunnen about 66.9% of the population (between age 25-64) have completed either non-mandatory upper secondary education or additional higher education (either University or a *Fachhochschule*).

Lauterbrunnen has an unemployment rate of 3.57%. As of 2005, there were 186 people employed in the primary economic sector and about 64 businesses involved in this sector. 197 people are employed in the secondary sector and there are 32 businesses in this sector. 1557 people are employed in the tertiary sector, with 194 businesses in this sector.[7] The historical population is given in the following table:[4]

year	population
1764	828
1850	1,756
1900	2,551
1910	3,204
1920	2,593
1950	2,876
1980	3,077[A]
2000	2,914

A Since 1973 includes Isenfluh

Origin of the name

According to locals, the name Lauterbrunnen is a combination of *lauter* meaning many, and *brunnen* meaning spring, fountain, or well. However, there is considerable dispute about the meaning of 'lauter', with some translating it as louder and others as clear, bright, or clean.

Transport

Lauterbrunnen railway station

The Berner Oberland Bahn (BOB) train runs to Interlaken.

The Wengernalpbahn (WAB) train leads to Kleine Scheidegg and on to Grindelwald,

The cable car and connecting train, both operated by the Bergbahn Lauterbrunnen-Mürren (BLM), provide service to Mürren. An alternative route to Mürren is available using the bus via the Trummelbach Falls to Stechelberg and then the Luftseilbahn Stechelberg-Mürren-Schilthorn (LSMS).

In other media

Johann Wolfgang von Goethe's poem *Gesang der Geister über den Wassern* (literal translation: Song of the Spirits above the Waters) was written while he stayed at the parish house near the Staubbach Falls waterfall in Lauterbrunnen. The Lauterbrunnen valley also provided the pictorial model for J. R. R. Tolkien's sketches and watercolours of the fictitious valley of Rivendell, and possibly also the name of the Bruinen river (meaning 'Loudwater') which flowed through it.

Lauterbrunnen featured in several scenes from the 1969 James Bond film *On Her Majesty's Secret Service*, including a car chase in which Bond (played for the only time by George Lazenby) was driven away from henchmen of Ernst Stavro Blofeld by his girlfriend Tracy di Vicenzo in a dramatic pursuit which culminated in them shaking off the pursuers in a stock car race.[8]

References

[1] Swiss Federal Statistical Office (http://www.bfs.admin.ch/bfs/portal/de/index/themen/01/02/blank/data/01.html), MS Excel document – *Bilanz der ständigen Wohnbevölkerung nach Kantonen, Bezirken und Gemeinden* (German) accessed 25 August 2010

[2] http://www.lauterbrunnen.ch

[3] http://www.bfs.admin.ch/bfs/portal/en/index/regionen/regionalportraets/gemeindesuche.html?geographyID=0584&FormEncoding=UTF-8

[4] *Lauterbrunnen* in German (http://www.hls-dhs-dss.ch/textes/d/D338.php), French (http://www.hls-dhs-dss.ch/textes/f/F338.php) and Italian (http://www.hls-dhs-dss.ch/textes/i/I338.php) in the online *Historical Dictionary of Switzerland*.

[5] Swiss Federal Statistical Office-Land Use Statistics (http://www.bfs.admin.ch/bfs/portal/de/index/themen/02/03/blank/data/gemeindedaten.html) 2009 data (German) accessed 25 March 2010

[6] John Ball, *The Alpine guide, Central Alps*, p. 75, 1866, London

[7] Swiss Federal Statistical Office (http://www.bfs.admin.ch/bfs/portal/en/index/regionen/regionalportraets/gemeindesuche.html) accessed 01-Jul-2009

[8] http://www.imdb.com/title/tt0064757/locations

External links

- www.lauterbrunnen.ch (http://www.lauterbrunnen.ch) Official website
- Anne-Marie Dubler: *Lauterbrunnen* in German (http://www.hls-dhs-dss.ch/textes/d/D338.php), French (http://www.hls-dhs-dss.ch/textes/f/F338.php) and Italian (http://www.hls-dhs-dss.ch/textes/i/I338.php) in the online *Historical Dictionary of Switzerland*.
- Truemmelbachfaelle (http://www.truemmelbachfaelle.ch)

Interlaken Ost railway station

Interlaken Ost regional rail	
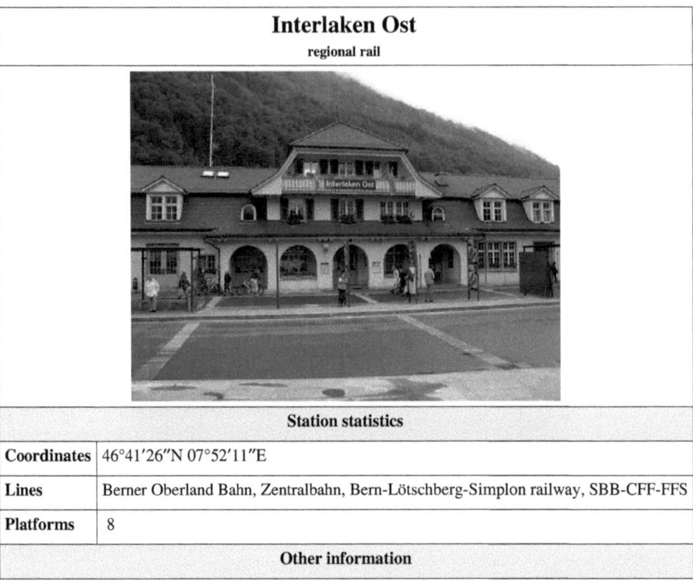	
Station statistics	
Coordinates	46°41′26″N 07°52′11″E
Lines	Berner Oberland Bahn, Zentralbahn, Bern-Lötschberg-Simplon railway, SBB-CFF-FFS
Platforms	8
Other information	

Interlaken Ost (East) is the main station in the town of Interlaken in the region of Bern, Switzerland.

Interlaken Ost is the main station of Interlaken, but there is also another station, Interlaken West.

Interlaken Ost is the terminus of various train services. The station is served by various different operators, these are: Berner Oberland Bahn, Zentralbahn, BLS, SBB-CFF-FFS and Deutsche Bahn.

A BOB train at Interlaken, note the former brown / cream livery

Train Services

The following services currently call at Interlaken Ost:

Operator	Train Type	Route	Frequency	Material	Notes
SBB	Intercity	Basel SBB - Olten - Bern - Thun - Spiez - Interlaken West - Interlaken Ost	1x per hour		
BLS	InterRegio	Bern - Thun - Spiez - Interlaken West - Interlaken Ost			
BLS	Regio Express	Zweisimmen - Boltigen - Erlenbach im Simmental - Oey-Diemtigen - Wimmis - Spiez - Interlaken West - Interlaken Ost			Golden Pass Panorama train
BLS	Regio	Spiez - Faulensee - Leissigen - Därligen - Interlaken West - Interlaken Ost	1x per hour		
Deutsche Bahn	ICE	Basel SBB - Liestal - Olten - Bern - Thun - Spiez - Interlaken West - Interlaken Ost			
Zentralbahn	Interregio (Brünig Line)	Interlaken Ost - Brienz - Meiringen - Brünig-Hasliberg - Lungern - Giswil - Sarnen - Alpnach Dorf - Hergiswil - Luzern	1x per hour		
Zentralbahn	Regio (Brünig Line)	Interlaken Ost - Riggenberg - Niederried - Oberried am Brienzersee - Ebligen - Brienz West - Brienz - Brienzwiler - Meiringen	1x per hour		
Berner Oberland Bahn	Regio	Interlaken Ost - Wilderswil - Zweilütschinen - Lauterbrunnen	2x per hour		
Berner Oberland Bahn	Regio	Interlaken Ost - Wilderswil - Zweilütschinen - Lütschental - Burglaunen - Schwendi - Grindelwald	2x per hour		

Preceding station	SBB-CFF-FFS	Following station
Interlaken West *toward Basel SBB*	SBB-CFF-FFS	*Terminus*
Interlaken West *toward Bern*	BLS	*Terminus*
Interlaken West *toward Spiez*	BLS	*Terminus*
Interlaken West *toward Zweisimmen*	BLS	*Terminus*
Terminus	BOB	Wilderswil *toward Lauterbrunnen*
Terminus	BOB	Wilderswil *toward Grindelwald*
Terminus	Zentralbahn	Brienz *toward Luzern*
Terminus	Zentralbahn	Riggenberg *toward Meiringen*
Interlaken West *toward Basel SBB*	1. default:ICE	*Terminus*

Zweilütschinen railway station

Zweilütschinen regional rail	
Station statistics	
Coordinates	46°38′N 7°54′E
Lines	Berner Oberland Bahn
Platforms	2
Other information	
Opened	1 July 1890
Electrified	17 March 1914

Zweilütschinen railway station serves the village of Zweilütschinen in the Swiss canton of Bern.

Next station South: **Sandweid**	Berner Oberland Bahn	Next station North: **Wilderswil**
Next station East: **Lutschental**	Berner Oberland Bahn	Next station North: **Wilderswil**

Sandweid railway station

Sandweid regional rail	
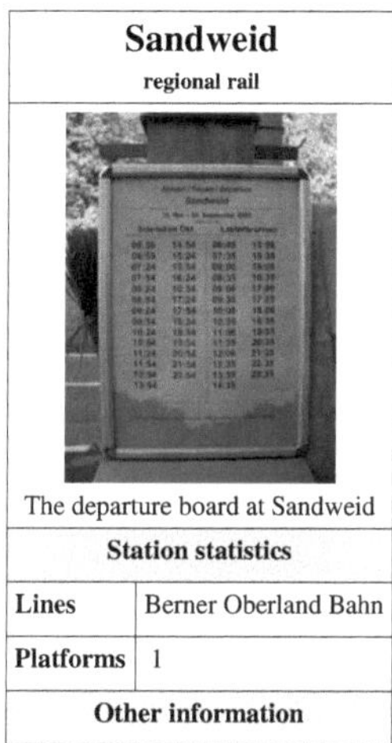 The departure board at Sandweid	
Station statistics	
Lines	Berner Oberland Bahn
Platforms	1
Other information	

Sandweid railway station serves the village of Sandweid in the Swiss canton of Bern.

It is a request stop station, and is not included in the printed public timetable produced by the Jungfrau railways company. A printed timetable is posted at the station.

Next station South: **Lauterbrunnen**	Berner Oberland Bahn	Next station North: **Zweilütschinen**

Lauterbrunnen railway station

Lauterbrunnen regional rail	
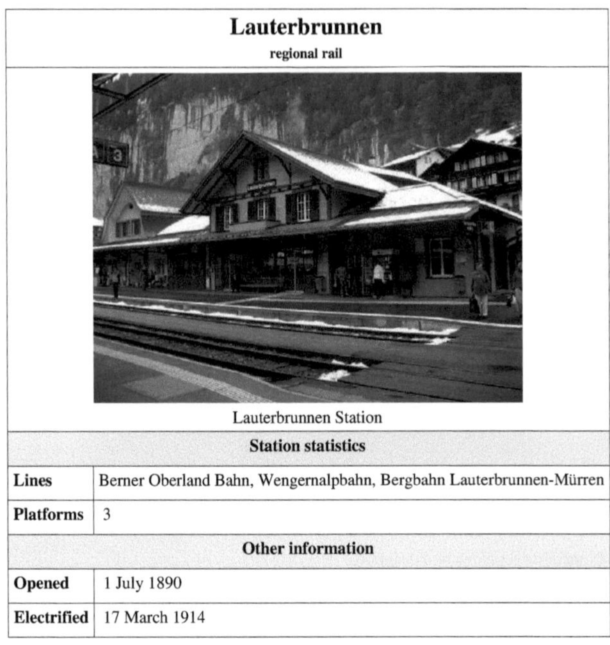 Lauterbrunnen Station	
Station statistics	
Lines	Berner Oberland Bahn, Wengernalpbahn, Bergbahn Lauterbrunnen-Mürren
Platforms	3
Other information	
Opened	1 July 1890
Electrified	17 March 1914

Lauterbrunnen railway station serves the village of Lauterbrunnen in the Swiss canton of Bern.

Connections

It has connections to the Wengernalpbahn for Kleine Scheidegg via Wengen and the Bergbahn Lauterbrunnen-Mürren to Mürren.

See also

- Jungfraubahn
- Grütschalp
- Winteregg
- Mürren railway station

Next station South: **Terminus**	Berner Oberland Bahn	Next station North: **Sandweid**

Next station South: **Terminus**	Wengernalpbahn	Next station North: **Wengwald**

Next station South: **Terminus**	Bergbahn Lauterbrunnen-Mürren	Next station North: **Grütschalp**

Lütschental railway station

Lutschental **regional rail**	
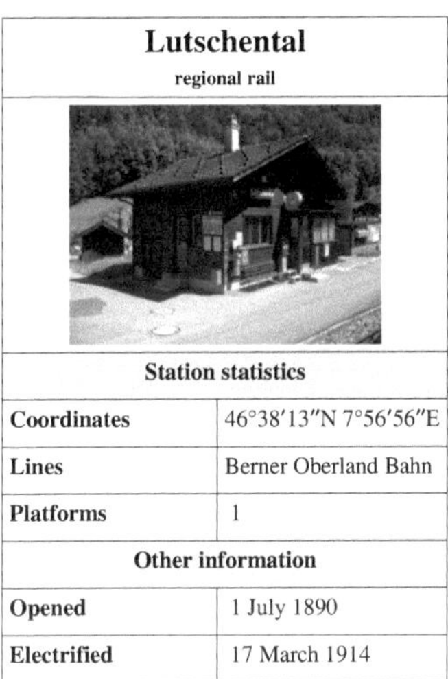	
Station statistics	
Coordinates	46°38′13″N 7°56′56″E
Lines	Berner Oberland Bahn
Platforms	1
Other information	
Opened	1 July 1890
Electrified	17 March 1914

Lütschental railway station serves the village of Lütschental in the canton of Berne, Switzerland.

Next station East: **Burglaunen**	Berner Oberland Bahn	Next station West: **Zweilütschinen**

Burglaunen railway station

Burglaunen regional rail	
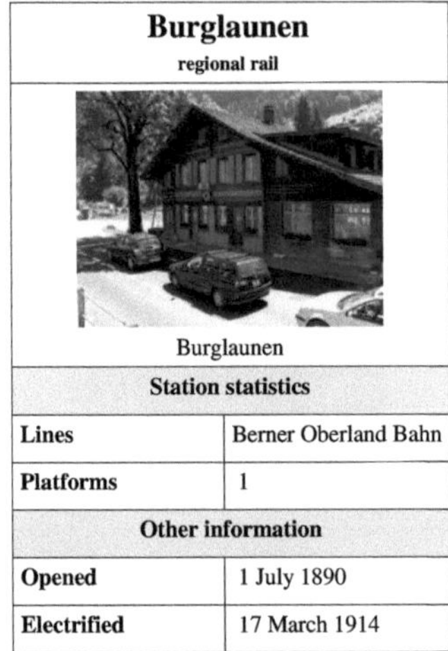 Burglaunen	
Station statistics	
Lines	Berner Oberland Bahn
Platforms	1
Other information	
Opened	1 July 1890
Electrified	17 March 1914

Burglaunen railway station serves the village of Burglaunen in the Swiss canton of Berne.

Next station East: **Schwendi**	Berner Oberland Bahn	Next station West: **Lutschental**

Jungfraubahn

Jungfraubahn	
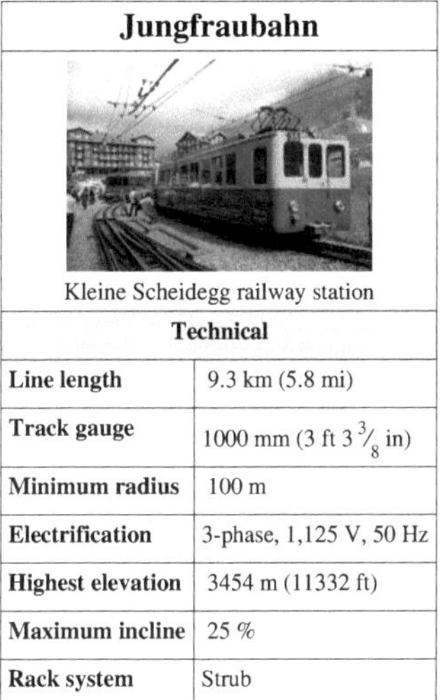 Kleine Scheidegg railway station	
Technical	
Line length	9.3 km (5.8 mi)
Track gauge	1000 mm (3 ft 3 3⁄8 in)
Minimum radius	100 m
Electrification	3-phase, 1,125 V, 50 Hz
Highest elevation	3454 m (11332 ft)
Maximum incline	25 %
Rack system	Strub

The **Jungfraubahn** (JB) is an 1000 mm (3 ft 3 3⁄8 in) gauge rack railway electrified at 3-phase 1,125 volts 50 Hertz, which runs 9 kilometres from Kleine Scheidegg to the highest railway station in Europe at Jungfraujoch. The railway runs almost entirely within a tunnel built into the Eiger and Mönch mountains and contains two stations in the middle of the tunnel, where passengers can disembark to observe the neighbouring mountains through windows built into the mountainside.

The JB is under the management of the Jungfraubahn Holding company, which also comprises the Wengener Alp Bahn (WAB) which links to the JB at Kleine Scheidegg and has two routes down the mountain, to Lauterbrunnen and Grindelwald, from where the Berner Oberland Bahn (BOB) connects to the Federal Railways at Interlaken.

History

- 1860 (approximately) - there were many different plans for a mountain railway on the Jungfrau, which failed due to financial problems.
- 1894 the industrialist Adolf Guyer-Zeller received a concession for a rack railway, which began from the Kleine Scheidegg railway station of the Wengernalpbahn (WAB), with a long tunnel through the Eiger and Mönch up to the summit of the Jungfrau.
- 1896 construction began. The construction work proceeded briskly.
- 1898 the Jungfraubahn opened as far as the Eigergletscher railway station, at the foot of the Eiger.
- 1899 Six workers are killed in an explosion. There is a 4 month strike by workers. Adolf Guyer-Zeller dies in Zürich on 3 April. The section from Eigergletscher to Rotstock opens on 2 August
- 1903 The section from Rotstock to Eigernordwand railway station opens on 28 June.
- 1905 The section from Eigernordwand to Eismeer railway station opens on 25 July

Construction of the Jungfraubahn

- 1908 There is an explosion at Eigernordwand railway station station.
- 1912 21 February sixteen years after works commenced the tunneling crew finally broke through the glacier in Jungfraujoch. The station at Jungfraujoch was inaugurated on 1 August.
- 1924 The Jungfraujoch house "The house above the clouds" is opened on 14 September.
- 1931 The research station at the Jungfraujoch is opened.
- 1937 The Sphinx Observatory is opened. A snowblower is purchased and this results in year-round operation.
- 1942 Relocation of the company offices from Zürich to Interlaken.
- 1950 The dome is installed on the Sphinx Observatory.
- 1951 The adhesion section between Eismeer railway station and Jungfraujoch railway station is converted to rack operation.
- 1955 A second depot at Kleine Scheidegg is constructed. The post office inaugurates its relay station on the Jungfraujoch.
- 1972 The panoramic windows are installed at Eigerwand and Eismeer. The Jungfraujoch mountain house and tourist house are destroyed by fire on 21 October.

Map of the Jungfraubahn project in 1903

- 1975 A new tourist house is opened.
- 1987 A new mountain house is opened on 1 August.
- 1991 A new station hall is opened at the Jungfraujoch.
- 1993 The small Kleine Scheidegg depot is extended.
- 1996 The covered observation deck at the Sphinx Observatory is opened.
- 1997 For the first time the annual visitor numbers exceed 500,000.
- 2000 On 1 June a daily record number of 8,148 visitors is received.

Stations

- Kleine Scheidegg, 2061 m (6762 ft)
- Eigergletscher, 2320 m (7612 ft)
- Eigernordwand, 2864 m (9396 ft)
- Eismeer, 3158 m (10361 ft)
- Jungfraujoch, 3454 m (11332 ft)

[1]

The Jungfraujoch railway station at 3,454 m above sea level

2008 Proposals

In early 2008, Jungfraubahn Holding AG announced it is exploring the futuristic idea of an efficient fast form of access to the Jungfraujoch as an addition to the rack railway. A feasibility study has been commissioned. The additional access would be the world's longest tunnel-lift system. The study is to show if and how such a tunnel-lift system - for example as a fast lift or funicular - from the Lauterbrunnen Valley to the Jungfraujoch could be realised without disturbing the unique landscape of the UNESCO World Heritage site.

The attractiveness of the cogwheel railway should thus be enhanced, as guests could use the fast lift for the uphill or downhill journey. Through a marked reduction in travelling time, the trip to the Jungfraujoch could also become a half-day excursion.

Rolling stock

Since most of the railway is inside a tunnel, it was designed to run with electricity from conception. The latest rolling stock consists of twin-unit motorcoaches carrying 230 per train which operate at 12.5 km/h on the steepest parts of the ascent. The motors function at two speeds which allows the units to operate at double this speed on the less steep part of the ascent (above Eismeer station).

The motors will operate in a regenerative mode which allows the trains to generate electricity during the descent which is fed back into the power distribution system. Approximately 50% of the energy required for an ascent is recovered during the descent. It is this generation that regulates the descent speed.

Snow clearing equipment is essential on the open section of line between Kleine Scheidegg railway station and Eigergletscher railway station. Originally snow ploughs were used but more recently snow blowing equipment has been brought into service.

The railway also operates some dedicated freight vehicles to supply the visitor facilities at Jungfraujoch, including a tank to transport additional water.

Main Characteristics

Altitude of top station above Sea Level	3454 m
Difference in height	1393 m
Operational length	9.3 km
Gauge	1000 mm (3 ft 3 $^{3}/_{8}$ in)
Rack rail type	Strub
Operational Speed	12.5 km/h (25 km/h on shallower gradients such as above Eismeer)
Steepest gradient	25%
Smallest curve radius	100 m
Tunnels	**3: longest 7122 m, shortest 110 m. 80% of length of the entire railway.**
Power system	**3-phase 50Hz 1125Volt**

Gallery

The Schreckhorn dominates the view from the window at the Eismeer station. One of two stations in the tunnel on the way to the Jungfraujoch

A snowblower at Kleine Scheidegg railway station.

The Jungfraubahn runs using a 1125 Volts three-phase alternating current system which requires the trains to collect power from twin overhead wires, using two pantographs, as seen here (the third phase is earthed to the track).

The strub rack system underneath a railcar.(Rowan locomotive He 2/2 no. 6)

A cogwheel from a Jungfrau Railway railcar strub rack system

See also

- Mountain railway
- Rail transport in Switzerland

References

[1] Between Heaven and Earth. History and technology - science and research on the Jungfraujoch - Top of Europe, Jungfraubahnen.

External links

- Jungfrau Railways website (http://www.jungfraubahn.ch/en/DesktopDefault.aspx/tabid-1/) (**English**)
- Video of a day trip to Jungfraubahn (http://video.google.com/videoplay?docid=6778419425314931660)

Schynige Platte Railway

Schynige Platte Bahn	
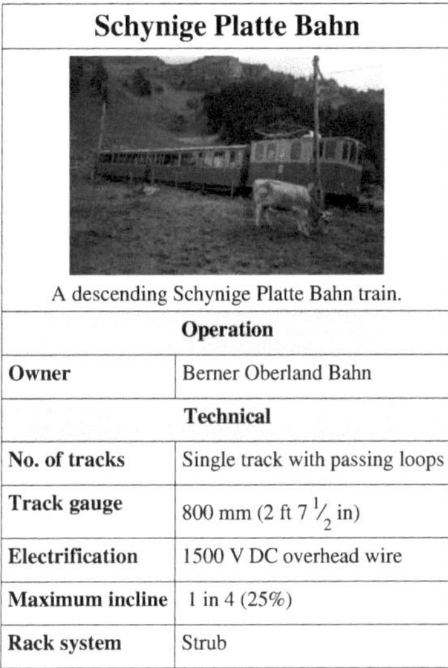 A descending Schynige Platte Bahn train.	
Operation	
Owner	Berner Oberland Bahn
Technical	
No. of tracks	Single track with passing loops
Track gauge	800 mm (2 ft 7 $^1/_2$ in)
Electrification	1500 V DC overhead wire
Maximum incline	1 in 4 (25%)
Rack system	Strub

The **Schynige Platte Railway** or **Schynige Platte Bahn** (SPB) is a mountain railway in the Bernese Oberland area of Switzerland, which connects the town of Wilderswil, near Interlaken with the famous wildflower gardens of the Schynige Platte.

An impressively varied natural landscape unfolds on the journey: fertile forests, Alpine pastures and breathtaking views of the Bernese Oberland with Lake Thun and Lake Brienz. Then a sudden change of scenery to the almost overwhelming view of the glistening giants of the Bernese Oberland, directly opposite, the imposing peaks of the Eiger, Mönch & Jungfrau.

History

- 1890 The concession for the line was given and the company is founded on 16 September.
- 1891 Construction starts.
- 1893 A train with special guests reaches the top on 5 May and the line opens to the public on 14 June.
- 1896 The Berner Oberland Bahn become the new owners.
- 1914 The line was electrified. Electric services start on 9 May.
- 1928 The Alpine Garden opens on the Schynige Platte.
- 1964 The Schynige Platte Railway receives 4 locomotives from the Wengernalpbahn.
- 1970 The Schynige Platte Railway receives 2 more locomotives from the Wengernalpbahn.
- 1978 The Schynige Platte Railway receives another locomotive from the Wengernalpbahn.
- 2001 Teddyland opens on the Schynige Platte on 7 June.

Operations

The Schynige Platte Railway connects with the metre gauge line of its owner company Berner Oberland Bahn (BOB) at Wilderswil. The line is 7.25 km (4.50 mi) long and has a rail gauge of 800 mm (2 ft 7 $^{1}/_{2}$ in). It is a rack railway, using the Strub rack to climb 1420 m (4659 ft) to its summit with a maximum gradient of 25% (1 in 4). There are a few short stretches of Riggenbach rack particularly over bridges.

Modern steel coach with owner inscription BOB B 45

Passage of the line is often be impossible in winter, and as there is no winter sports possibility, it only operates in summer months, generally from the end of April to the end October.

The line is electrified using a 1500 V DC overhead supply since 1914. The line still operates the four original electric locomotives, numbers 11 to 14. More such locomotives, together with matching coaches, were bought from the Wengernalpbahn but had (until the recent delivery of additional train sets) often returned to this line to help in winter sports traffic. 1992 the company started a rebuilding program for the coaches. On old underframes, brought to a unified length of 10,695 mm over buffers, Ramseyer+Jenzer built a new steel body in the old compartment style.

No. 14, one of the original locomotives in heritage livery in Wilderswil station

BOB and all companies of the Jungfraubahn Holding (JBH) have a joint management, staff may be exchanged and technical services are available for other companies within the group.

Locomotives

No.	Name	Arms	Builder	Date	SPB	Livery (2008)	Notes
5			SLM	1894	1894-		Class H2/3, 0-4-2T, Steam.
11	*Wilderswil*		SLM / BBC	1914	1914-		All electric locomotives, Class He2/2
12	*(Gsteigwiler)*		SLM/BBC	1914	1914-		
13	*Matten*		SLM/BBC	1914	1914-		
14	*(Gündlischwand)*		SLM/BBC	1914	1914-		
15			SLM/Alioth	1910	1964-92		originally WAB 55, 1992 back to WAB as shunter Lauterbrunnen, 1997 monument in Münchenstein BL (former Alioth factory)
16	*Anemone*		SLM/Alioth	1910	1964-		originally WAB 56
17			SLM/Alioth	1910	1964-96		originally WAB 57, withdrawn 1996

18	*(Krokus)* *Gündlischwand*		SLM/Alioth	1910	1964-		originally WAB 58
19	*Flühbluhme*		SLM/Alioth	1911	1964-		originally WAB 59
20	*(Edelweiss)* *Gsteigwiler*		SLM/Alioth	1911	1970-		originally WAB 60
21 61	*Enzian*		SLM/Alioth	1912	1970-81 1991		Rebuilt 1992
62	*Alpenrose*		SLM/Alioth	1912	1989-		Rebuilt 1989
63	*(Silberdistel)*		SLM/Alioth	1912	1996-		Rebuilt 1996

References

- Brawand Hansruedi, *Schynige Platte Bahn, Die Bergstrecke der Berner Oberland-Bahnen, Mit nostalgischem Cachet.* Prellbock Verlag, Leissigen, 2003, ISBN 3-907579-26-7
- Book *Tramways and Light Railways of Switzerland and Austria*, ISBN 0-900433-96-5, by R.J.Buckley, published by the Light Rail Transit Association, 1984.
- Rolling stock lists by Verein Rollmaterialverzeichnis Schweiz

External links

- Schynige Platte Bahn website (http://www.jungfrau.ch/en/tourism/places-to-visit/schynige-platte/)
- Some Photographs and a brief description (http://www.strollingguides.co.uk/books/berneroberland/places/spb.php)

Bergbahn Lauterbrunnen–Mürren

Bergbahn Lauterbrunnen-Mürren	
 Two trains pass in Winteregg station	
Technical	
Line length	4.2 km (2.6 mi)
No. of tracks	Single track with one passing loop
Track gauge	1000 mm (3 ft 3 ³⁄₈ in)
Electrification	525 V DC
Maximum incline	5 %
Rack system	None

The **Bergbahn Lauterbrunnen-Mürren** (BLM) is a mountain railway in the Bernese Oberland area of Switzerland, which connects the villages of Lauterbrunnen and Mürren.

The former funicular above Lauterbrunnen

Operations

The enterprise consists of an aerial cableway and a mountain railway, connecting at Grütschalp station. The cableway runs 1.4 km (0.9 mi) from Lauterbrunnen to Grütschalp, and was opened only in 2006, replacing a funicular with maximum gradient of 60%. From Grütschalp to Mürren the line is continued as a 4.2 km (2.6 mi) long narrow gauge electric railway with a maximum upward gradient of 5% and using adhesion traction (i.e. not rack and pinion). The railway is single-track, with a passing loop at Winteregg station.

Both the funicular and electric railway had a track gauge of 1000 mm (3 ft 3 ³⁄₈ in), enabling the use of the funicular to transfer rolling stock to and from the electric railway. A travelling crane in the Grütschalp station is used to transfer goods between the two sections - well used since road access to Mürren is very difficult. With the replacement of the funicular, the aerial cable car has also taken over the task of transporting goods. The single cable car carries up to 100 passengers in a journey time of 4 minutes, a significant improvement over the 11 minute funicular journey time. The special crane has been conserved to facilitate transfer of goods between the railway and a goods platform slung underneath the aerial cablecar.

The approach to Murren

The BLM station at Lauterbrunnen is directly opposite the station of the Berner Oberland Bahn (BOB) to Interlaken, and the Wengernalpbahn (WAB) to Kleine Scheidegg and Grindelwald.

History

- 1887 Concession obtained for the construction of the railways.
- 1889 The company is formed and construction starts.
- 1891 Railway opens. The planned opening on 1 June is delayed until 14 August due to a derailment.
- 1902 The funicular railway is converted from water gravity power to electric power .
- 1910 First winter operations started.
- 1912 Replacement of the locomotives on the Mürren to Grütschalp section by motor coaches (type BDe 2/4).
- 1949 New vehicles and rope are installed on the Lauterbrunnen to Grütschalp section.
- 1965 The new station at Mürren is opened.
- 1994 The freight loading operations at Grütschalp are rebuilt.
- 2006 Last operation of the funicular railway from Lauterbrunnen to Grütschalp is on 23 April and the first operation of the replacement cablecar was on 16 December.

References

- Buckley, R.J. (1984). *Tramways and Light Railways of Switzerland and Austria*. Light Rail Transit Association. ISBN 0900433965..

External links

- Bergbahn Lauterbrunnen-Mürren [1] in English. Part of a site about the Swiss narrow gauge railways.

References

[1] http://www.rail-info.ch/BLM/index.en.html

Article Sources and Contributors

Grindelwald railway station *Source*: http://en.wikipedia.org/w/index.php?title=Grindelwald_railway_station *Contributors*: Bahnfrend, Cacolantern, D6, Waacstats, 4 anonymous edits

Grindelwald *Source*: http://en.wikipedia.org/w/index.php?title=Grindelwald *Contributors*: Ajho, Aliman5040, AllyUnion, Aranel, Attilios, BillC, Bryan Derksen, Chatfecter, Chochopk, Ckatz, Cometstyles, Cruccone, D6, David Shay, De728631, DidiWeidmann, Docu, Ekem, Ericoides, Fairchildd, Glacier109, Halsteadk, Hapesoft, Hhk1989, Ikiwaner, Irmgard, Japanese Searobin, Jdhowens90, Julien29, Jwissick, Kcarnold, Kenneth Stephen, Knockmoreguy96, Lajsikonik, MadGeographer, Malloryrdc, MichaelBillington, Naddy, Ncsakany, Nicolas1981, Nniuq25, Olivier, Osarius, Pp2007, Rama, RedWolf, Rich Farmbrough, Richard1966, Sabinpopa, Skew-t, Stan Shebs, Steinninn, Ted Wilkes, Thisisbossi, Thorwald, Timfpark, Tobyc75, TonyW, TracksOnWax, Tudorminstrel, Wilchett, Xezbeth, Zacharie Grossen, 35 anonymous edits

Bern *Source*: http://en.wikipedia.org/w/index.php?title=Bern *Contributors*: .:Ajvol:., 16@r, 3122WIKI, 334a, 62.2.17.xxx, A-research, Adam Carr, Adam Krellenstein, Adebaumann, Aervanath, Ahoerstemeier, Aitias, Ajnem, Alan.ca, Alansohn, Amalthea, Ami in CH, Anarcho punk1, Anaxial, Andre Engels, Andrewpmk, Andrwsc, Angr, AniBunny, Apterygial, ArRoos, Atlant, Attilios, Audionaut, Barefeetdude, Bejnar, Ben-Zin, Benjaminevans82, BillC, Biruitorul, Blehfu, Bobak, Bobfrombrockley, Bojin, Bokpasa, Bubba hotep, CLW, CalicoCatLover, Calmer Waters, CambridgeBayWeather, Caponer, Cassowary, Cautious, Cflm001, Chanheigeorge, CharlesC, Chase22134, Chaz1dave, Chillwills, Chris the speller, Chrisch, Christian List, Clpda, Cmdrjameson, Coder17, Cold starlight, Conversion script, Coolcaesar, Coyets, Cpcallen, Cs-wolves, Csvndl4, D6, Daboss, Danny, Dbachmann, Dcwoon, De la mettrie, Deacon of Pndapetzim, Defrenrokorit, DerBorg, DerRichter, DerrickOswald, Dewritech, DisillusionedBitterAndKnackered, Dmn, Docu, Douwe20, Dr. Blofeld, Dschwen, Eastfrisian, Edcolins, Egil, El C, Enviroboy, Everyking, Excirial, Fak119, Fireaxe888, Forteblast, Funnybunny, Funnyhat, Furrykef, GC, Gabbe, Gaius Cornelius, Galoubet, Gentschli, Georgequizbowl08, Glaurung, Globi2002, Godardesque, Goldfishbutt, Graham87, Great Deku Tree, Greenshed, Groshna, Grstain, Gwernol, Hashar, Hayden120, Helohe, Hermes Agathos, Hmains, Hughcharlesparker, Hyperfusion, Hyperplane, If62668, Ikiwaner, Infrogmation, Iokseng, J. 'mach' wust, JREL, Jahanas, Jeppiz, Jeronimo, Jesuit222, Jezerfetnae, Jhendin, JinJian, Jinglebells200, JoanneB, Joey80, Jose77, Joshua, KAVEBEAR, KFP, Kaaveh Ahangar, KaiserAO, Kcuello, Kerotan, Knatterton, Knutux, Korte, Kotasik, Kwaku, Kwamikagami, Kyle1278, Lafuzion, Larsbaumgartner, Layoub85, Le Fou, Leftydan6, LeoNomis, Leyo, Lfh, Lightmouse, Longbow4u, Lukas Diener, Lupin, MacTire02, MadGeographer, Maksim L., Marek69, Mark Wheaver, Marquez, Mathpianist93, Mav, Maximus Rex, Maya, Mbz1, MeltBanana, Michael Jackson (not king of pop), Michellecrisp, Mild Bill Hiccup, Millisits, Momet, Mschlindwein, N panni, N-true, Nakon, Natalie Erin, NawlinWiki, Newnoise, Ngagnebin, Nguyen Thanh Quang, Nico, Nova77, Odryfuss, Ojw, Olivier, OnBeyondZebrax, Optimale, Ori, OwenBlacker, Paulie74, Pcaldwell76, Pedant17, Peter S., Phil Bastian, Photnart, Pinethicket, Pitsteelerfan188, Pmanderson, Pne, Postlebury, Pras, Premierathon, President Rhapsody, PsY.cHo, Pully992, Quadell, Queenofthewilis, Raphael Frey, Red Winged Duck, RedWolf, Reisio, RexNL, Reywas92, RicciSpeziari, Rich Farmbrough, RickK, Rl, Robofish, Rocastelo, Roidhrigh, SAMbo, Sailsbystars, Salt Yeung, Sam Hocevar, Sandstein, Saper, Sardanaphalus, Saxsux, Scipius, Scotthatton, Sdnegel, Sherool, Signalhead, SimonP, Sir Stanley, Skew-t, Skinsmoke, Slady, Sluzzelin, Smalljim, SnapSnap, Snowolf, Sowen, Spigot, Spondoolicks, Ste, Stewartadcock, Sweden555, Synthetik, TGC55, TPIRman, TRBlom, Tabletop, Thisisbossi, ThuranX, Tian, Timwi, Tobias Hoevekamp, Tobyc75, Tomeasy, Trilobite, Truthkeeper88, Twenty-nine, Universalcosmos, Van Peter, Van helsing, Vanished user 03, Verrai, WTell, Wathiik, WereSpielChequers, WhisperToMe, Wik, Wikibob, Wikky Horse, WildWildBil, Wimt, Wmem, Woohookitty, Worldphotopage, Writtenright, Ww2censor, Yidisheryid, Zacharie Grossen, Zeledi, Zenit, Zscout370, Ég er Almar, Île flottante, Александър, , 424 anonymous edits

Grindelwald Grund railway station *Source*: http://en.wikipedia.org/w/index.php?title=Grindelwald_Grund_railway_station *Contributors*: Bahnfrend, D6, Waacstats, Wwoods, 2 anonymous edits

Schwendi railway station *Source*: http://en.wikipedia.org/w/index.php?title=Schwendi_railway_station *Contributors*: Bahnfrend, Cacolantern, D6, Docu, Waacstats, 1 anonymous edits

Wengernalpbahn *Source*: http://en.wikipedia.org/w/index.php?title=Wengernalpbahn *Contributors*: AHEMSLTD, AndrewHowse, Arwel Parry, Billyray32, Bryan Derksen, CComMack, Cacolantern, Chris j wood, Coyets, D6, Davehi1, DerBorg, Docu, G-Man, Gwernol, Halsteadk, Hapesoft, Henning Makholm, JaGa, Keilana, Osarius, Peter Horn, Peterjpreston, Reimgild, Rich Farmbrough, RonaldHummelink, Shortfatlad, Skew-t, Synthetik, TrackConversion, TracksOnWax, Tudorminstrel, 44 anonymous edits

Berner Oberland Bahn *Source*: http://en.wikipedia.org/w/index.php?title=Berner_Oberland_Bahn *Contributors*: Ajho, Arwel Parry, Bermicourt, Bobblewik, CComMack, Cacolantern, Chris j wood, Danyg, Darkwind, DerBorg, Docu, Enotayokel, Gestumblindi, Grandpapas, Gwernol, Gürbetaler, Halsteadk, Henning Makholm, Imgaril, MadGeographer, Markussep, Mjroots, Osarius, Peter Horn, Pumpie, Rjwilmsi, Shortfatlad, Skew-t, SkinnyPrude, Slambo, Svick, Tiddly Tom, Tudorminstrel, Woohookitty, 30 anonymous edits

Schynige Platte *Source*: http://en.wikipedia.org/w/index.php?title=Schynige_Platte *Contributors*: Audriusa, ChrisJ, Docu, Droll, Hapesoft, Lombar2, Mrslippery, ZachT, Zacharie Grossen, 3 anonymous edits

Wilderswil *Source*: http://en.wikipedia.org/w/index.php?title=Wilderswil *Contributors*: Aliman5040, BillC, D6, MadGeographer, Marek69, Rayc, Rigadoun, Tobyc75, 1 anonymous edits

Wilderswil railway station *Source*: http://en.wikipedia.org/w/index.php?title=Wilderswil_railway_station *Contributors*: Bahnfrend, Cacolantern, D6, Dontheguy, Michaelas10, Rbraunwa, Stemonitis, Vanish2, Waacstats, 15 anonymous edits

Lauterbrunnen *Source*: http://en.wikipedia.org/w/index.php?title=Lauterbrunnen *Contributors*: Aliman5040, Aoi, Attilios, BillC, CBDunkerson, Chatfecter, Chris j wood, D6, Datrio, David Newton, Docu, EquatorialSky, Ericoides, Gestumblindi, Halsteadk, Hapesoft, Ikiwaner, Irmgard, J. 'mach' wust, Kate, Kenwbar, Linuxaurus, MadGeographer, Marine4, Michael Zimmermann, Nateedge, Ngchikit, Nilington, Olivier, Osarius, Rich Farmbrough, Roarman, Sir Stanley, Skew-t, Sluzzelin, Tassedethe, Tobyc75, Wikibob, Yesuitus2001, Zacharie Grossen, 35 anonymous edits

Interlaken Ost railway station *Source*: http://en.wikipedia.org/w/index.php?title=Interlaken_Ost_railway_station *Contributors*: Bahnfrend, Cacolantern, Chris0693, D6, DaSch, Dontheguy, Grahamec, Jonathan Jeremiah Peachum, Sw2nd, Vanish2, Waacstats, 15 anonymous edits

Zweilütschinen railway station *Source*: http://en.wikipedia.org/w/index.php?title=Zweil%C3%BCtschinen_railway_station *Contributors*: Bahnfrend, Cacolantern, Docu, Waacstats, 5 anonymous edits

Sandweid railway station *Source*: http://en.wikipedia.org/w/index.php?title=Sandweid_railway_station *Contributors*: Bahnfrend, Cacolantern, D6, Waacstats, 4 anonymous edits

Lauterbrunnen railway station *Source*: http://en.wikipedia.org/w/index.php?title=Lauterbrunnen_railway_station *Contributors*: Bahnfrend, Cacolantern, D6, DerBorg, Waacstats, 8 anonymous edits

Lütschental railway station *Source*: http://en.wikipedia.org/w/index.php?title=L%C3%BCtschental_railway_station *Contributors*: Bahnfrend, Cacolantern, Docu, Waacstats, 5 anonymous edits

Burglaunen railway station *Source*: http://en.wikipedia.org/w/index.php?title=Burglaunen_railway_station *Contributors*: Bahnfrend, Cacolantern, D6, Docu, Waacstats, 1 anonymous edits

Jungfraubahn *Source*: http://en.wikipedia.org/w/index.php?title=Jungfraubahn *Contributors*: Arwel Parry, Chris j wood, Cory Donnelly, Coyets, Crampon, Cryonic07, DerBorg, GregorB, Gwernol, Halsteadk, Henning Makholm, Iain Bell, Jensgram, Keimzelle, Keinstein, Kroschka Ru, Lightmouse, MadGeographer, Mu2, Oldie, Olivier, Osarius, Peter Horn, RHaworth, RattusMaximus, Samaster1991, Shortfatlad, Skew-t, Sladen, Stan Shebs, Synthetik, Tabletop, 30 anonymous edits

Schynige Platte Railway *Source*: http://en.wikipedia.org/w/index.php?title=Schynige_Platte_Railway *Contributors*: Ajho, Audriusa, Bantman, Chris j wood, DerBorg, Docu, Gaius Cornelius, Gwernol, Gürbetaler, Halsteadk, Hapesoft, Henning Makholm, MadGeographer, Michael Johnson, Moonraker12, Osarius, Peter Horn, Rich Farmbrough, SalomonCeb, Sandstein, Shortfatlad, Skew-t, Tabletop, TrackConversion, Tudorminstrel, 12 anonymous edits

Bergbahn Lauterbrunnen–Mürren *Source*: http://en.wikipedia.org/w/index.php?title=Bergbahn_Lauterbrunnen%E2%80%93M%C3%BCrren *Contributors*: Bobblewik, Brockert, CatherineMunro, Chris j wood, DerBorg, Docu, Enotayokel, Gestumblindi, Good Olfactory, Gwernol, Gürbetaler, Halsteadk, Harold, Henning Makholm, Iain Bell, Jachapo, Jashton56, Joseolgon, Konrad-EN, Osarius, Peter Horn, Peter Karlsen, Pumpie, RHaworth, Saltose, Svens Welt, Thrapper, 11 anonymous edits

Image Sources, Licenses and Contributors

File:Grindelwald train station.jpg *Source*: http://en.wikipedia.org/w/index.php?title=File:Grindelwald_train_station.jpg *License*: Public Domain *Contributors*: Ncsakany

File:Grindelwald Wetterhorn.jpg *Source*: http://en.wikipedia.org/w/index.php?title=File:Grindelwald_Wetterhorn.jpg *License*: Public Domain *Contributors*: Ncsakany

File:Grindelwald-coat of arms.png *Source*: http://en.wikipedia.org/w/index.php?title=File:Grindelwald-coat_of_arms.png *License*: GNU Free Documentation License *Contributors*: -

file:Switzerland location map.svg *Source*: http://en.wikipedia.org/w/index.php?title=File:Switzerland_location_map.svg *License*: Creative Commons Attribution-Sharealike 3.0 *Contributors*: Lukasb1992, NordNordWest, Sting, 1 anonymous edits

File:Red pog.svg *Source*: http://en.wikipedia.org/w/index.php?title=File:Red_pog.svg *License*: Public Domain *Contributors*: User:Andux

File:4992 - Grindelwald - Kirch am Gydisdorf.JPG *Source*: http://en.wikipedia.org/w/index.php?title=File:4992_-_Grindelwald_-_Kirch_am_Gydisdorf.JPG *License*: Attribution *Contributors*: User:Thisisbossi

File:Grindelwald.jpg *Source*: http://en.wikipedia.org/w/index.php?title=File:Grindelwald.jpg *License*: Public Domain *Contributors*: User:Jphoto

File:Grindelwaldviewfromhill.jpg *Source*: http://en.wikipedia.org/w/index.php?title=File:Grindelwaldviewfromhill.jpg *License*: Public Domain *Contributors*: User:Hhk1989

File:Bern luftaufnahme.png *Source*: http://en.wikipedia.org/w/index.php?title=File:Bern_luftaufnahme.png *License*: Public Domain *Contributors*: AndreasPraefcke, Docu, Robert Illes, Sandstein, Trockennasenaffe

File:Wappen Bern matt.svg *Source*: http://en.wikipedia.org/w/index.php?title=File:Wappen_Bern_matt.svg *License*: unknown *Contributors*: User:Sa-se

File:Speaker Icon.svg *Source*: http://en.wikipedia.org/w/index.php?title=File:Speaker_Icon.svg *License*: Public Domain *Contributors*: Blast, G.Hagedorn, Mobius, 2 anonymous edits

File:Untertorbrücke Tschachtlanchronik.jpg *Source*: http://en.wikipedia.org/w/index.php?title=File:Untertorbrücke_Tschachtlanchronik.jpg *License*: Public Domain *Contributors*: Bendict Tschachtlan

File:MerianBern.jpg *Source*: http://en.wikipedia.org/w/index.php?title=File:MerianBern.jpg *License*: Public Domain *Contributors*: Vorlage von Joseph Plepp

File:Aareschlaufe Bern-East.svg *Source*: http://en.wikipedia.org/w/index.php?title=File:Aareschlaufe_Bern-East.svg *License*: unknown *Contributors*: 3122WIKI, Docu

File:Erlacherhof.jpg *Source*: http://en.wikipedia.org/w/index.php?title=File:Erlacherhof.jpg *License*: Creative Commons Attribution 3.0 *Contributors*: User:Sandstein

File:Rathaus (Bern).jpg *Source*: http://en.wikipedia.org/w/index.php?title=File:Rathaus_(Bern).jpg *License*: GNU Free Documentation License *Contributors*: Original uploader was Wladyslaw Sojka at de.wikipedia (Original text : ? Wladyslaw [Disk.])

File:CH Bern Kramgasse.jpg *Source*: http://en.wikipedia.org/w/index.php?title=File:CH_Bern_Kramgasse.jpg *License*: Creative Commons Attribution-Sharealike 2.5 *Contributors*: User:Dschwen

File:Zentrum Paul Klee Bern 15.JPG *Source*: http://en.wikipedia.org/w/index.php?title=File:Zentrum_Paul_Klee_Bern_15.JPG *License*: Creative Commons Attribution 2.5 *Contributors*: User:Noebu

File:StadttheaterBern.JPG *Source*: http://en.wikipedia.org/w/index.php?title=File:StadttheaterBern.JPG *License*: unknown *Contributors*: Photographed by Supermutz

File:Gurtenfestival Gelaende 2003.JPG *Source*: http://en.wikipedia.org/w/index.php?title=File:Gurtenfestival_Gelaende_2003.JPG *License*: Creative Commons Attribution 3.0 *Contributors*: User:Sprain

File:Stadedesuisse-2.jpg *Source*: http://en.wikipedia.org/w/index.php?title=File:Stadedesuisse-2.jpg *License*: GNU Free Documentation License *Contributors*: User:Hb-mfb

File:Combino Bern.jpg *Source*: http://en.wikipedia.org/w/index.php?title=File:Combino_Bern.jpg *License*: GNU Free Documentation License *Contributors*: Hauser Christoph

File:Einsteinhausberne.jpg *Source*: http://en.wikipedia.org/w/index.php?title=File:Einsteinhausberne.jpg *License*: Public Domain *Contributors*: User:Korte

File:Grund.JPG *Source*: http://en.wikipedia.org/w/index.php?title=File:Grund.JPG *License*: Public Domain *Contributors*: Original uploader was Andrewrabbott at en.wikipedia

File:BOB Schwendi Station.jpg *Source*: http://en.wikipedia.org/w/index.php?title=File:BOB_Schwendi_Station.jpg *License*: Creative Commons Attribution-Sharealike 2.5 *Contributors*: User:SalomonCeb

file:Mh kleine scheidegg sommer.jpeg *Source*: http://en.wikipedia.org/w/index.php?title=File:Mh_kleine_scheidegg_sommer.jpeg *License*: GNU Free Documentation License *Contributors*: user:LosHawlos

Image:GGDiamondCrossing.JPG *Source*: http://en.wikipedia.org/w/index.php?title=File:GGDiamondCrossing.JPG *License*: Public Domain *Contributors*: Original uploader was Andrewrabbott at en.wikipedia

Image:KleineScheideggTriangle.JPG *Source*: http://en.wikipedia.org/w/index.php?title=File:KleineScheideggTriangle.JPG *License*: Public Domain *Contributors*: Original uploader was Andrewrabbott at en.wikipedia

Image:WengenFreight.JPG *Source*: http://en.wikipedia.org/w/index.php?title=File:WengenFreight.JPG *License*: Public Domain *Contributors*: Original uploader was Andrewrabbott at en.wikipedia

file:Mh BOB Interlaken.jpeg *Source*: http://en.wikipedia.org/w/index.php?title=File:Mh_BOB_Interlaken.jpeg *License*: GNU Free Documentation License *Contributors*: User:LosHawlos

Image:BOBGrindlewald.JPG *Source*: http://en.wikipedia.org/w/index.php?title=File:BOBGrindlewald.JPG *License*: Public Domain *Contributors*: Original uploader was Andrewrabbott at en.wikipedia

File:Wilderswil20080629Y577 Bf BOB313 SPB50.jpg *Source*: http://en.wikipedia.org/w/index.php?title=File:Wilderswil20080629Y577_Bf_BOB313_SPB50.jpg *License*: unknown *Contributors*: User:Gürbetaler

Image:Sandweid.JPG *Source*: http://en.wikipedia.org/w/index.php?title=File:Sandweid.JPG *License*: Public Domain *Contributors*: Original uploader was Andrewrabbott at en.wikipedia

File:20051014S147 ABt422.jpg *Source*: http://en.wikipedia.org/w/index.php?title=File:20051014S147_ABt422.jpg *License*: unknown *Contributors*: User:Gürbetaler

Image:Schynige Platte.jpg *Source*: http://en.wikipedia.org/w/index.php?title=File:Schynige_Platte.jpg *License*: GNU Free Documentation License *Contributors*: User:Audriusa

File:Fire.svg *Source*: http://en.wikipedia.org/w/index.php?title=File:Fire.svg *License*: Public Domain *Contributors*: User:Indolences

File:WilderswilStation.JPG *Source*: http://en.wikipedia.org/w/index.php?title=File:WilderswilStation.JPG *License*: Public Domain *Contributors*: Original uploader was Andrewrabbott at en.wikipedia

File:Wilderswil-coat of arms.svg *Source*: http://en.wikipedia.org/w/index.php?title=File:Wilderswil-coat_of_arms.svg *License*: Creative Commons Attribution-Sharealike 2.5 *Contributors*: User:Aliman5040

File:Wilderswil.jpg *Source*: http://en.wikipedia.org/w/index.php?title=File:Wilderswil.jpg *License*: GNU Free Documentation License *Contributors*: Elisabeth Belik

File:Lauterbrunnen valley in summer.jpg *Source*: http://en.wikipedia.org/w/index.php?title=File:Lauterbrunnen_valley_in_summer.jpg *License*: Creative Commons Attribution-Sharealike 3.0 *Contributors*: Halsteadk (talk)

File:Lauterbrunnen-coat of arms.svg *Source*: http://en.wikipedia.org/w/index.php?title=File:Lauterbrunnen-coat_of_arms.svg *License*: Creative Commons Attribution-Sharealike 2.5 *Contributors*: User:Aliman5040

File:LauterbrunnenValley.jpg *Source*: http://en.wikipedia.org/w/index.php?title=File:LauterbrunnenValley.jpg *License*: Creative Commons Attribution-Sharealike 3.0 *Contributors*: User:Weasel707

Image:LautebrunnenValley earlyMay2007.jpg *Source*: http://en.wikipedia.org/w/index.php?title=File:LautebrunnenValley_earlyMay2007.jpg *License*: Public Domain *Contributors*: EquatorialSky

Image:LauterbrunnenStation.JPG *Source*: http://en.wikipedia.org/w/index.php?title=File:LauterbrunnenStation.JPG *License*: Public Domain *Contributors*: Original uploader was Andrewrabbott at en.wikipedia

File:BOB Interlaken Ost Station.jpg *Source*: http://en.wikipedia.org/w/index.php?title=File:BOB_Interlaken_Ost_Station.jpg *License*: Creative Commons Attribution-Sharealike 2.5 *Contributors*: User:SalomonCeb

Image:Mh BOB Interlaken.jpeg *Source*: http://en.wikipedia.org/w/index.php?title=File:Mh_BOB_Interlaken.jpeg *License*: GNU Free Documentation License *Contributors*: User:LosHawlos

File:BOB Zweiluetschinen Station.jpg *Source*: http://en.wikipedia.org/w/index.php?title=File:BOB_Zweiluetschinen_Station.jpg *License*: Creative Commons Attribution-Sharealike 2.5 *Contributors*: User:SalomonCeb

File:Sandweid.JPG *Source*: http://en.wikipedia.org/w/index.php?title=File:Sandweid.JPG *License*: Public Domain *Contributors*: Original uploader was Andrewrabbott at en.wikipedia

File:LauterbrunnenStation.JPG *Source*: http://en.wikipedia.org/w/index.php?title=File:LauterbrunnenStation.JPG *License*: Public Domain *Contributors*: Original uploader was Andrewrabbott at en.wikipedia

File:LutschentalStation.JPG *Source*: http://en.wikipedia.org/w/index.php?title=File:LutschentalStation.JPG *License*: Public Domain *Contributors*: Original uploader was Andrewrabbott at en.wikipedia

File:BOB Burglauenen Station.jpg *Source*: http://en.wikipedia.org/w/index.php?title=File:BOB_Burglauenen_Station.jpg *License*: Creative Commons Attribution-Sharealike 2.5 *Contributors*: User:SalomonCeb

file:Mh JB kleine scheidegg.jpeg *Source*: http://en.wikipedia.org/w/index.php?title=File:Mh_JB_kleine_scheidegg.jpeg *License*: GNU Free Documentation License *Contributors*: Gürbetaler, Juiced lemon, LosHawlos, ZorkNika

File:Jungfraubahn Bohrarbeiten um 1900 (01).jpg *Source*: http://en.wikipedia.org/w/index.php?title=File:Jungfraubahn_Bohrarbeiten_um_1900_(01).jpg *License*: Public Domain *Contributors*: User:Norbert Böber

File:Jungfraubahn Plan 1903.jpg *Source*: http://en.wikipedia.org/w/index.php?title=File:Jungfraubahn_Plan_1903.jpg *License*: Public Domain *Contributors*: Schweizerisches Bundesarchiv

File:Bahnhof Jungfraujoch.jpg *Source*: http://en.wikipedia.org/w/index.php?title=File:Bahnhof_Jungfraujoch.jpg *License*: Creative Commons Attribution-Sharealike 3.0 *Contributors*: Adrian Sulc Original uploader was Hinterkappelen at de.wikipedia

Image:Eismeer window.jpg *Source*: http://en.wikipedia.org/w/index.php?title=File:Eismeer_window.jpg *License*: unknown *Contributors*: User:Stan Shebs

Image:Jungfraubahn Snowplough.JPG *Source*: http://en.wikipedia.org/w/index.php?title=File:Jungfraubahn_Snowplough.JPG *License*: Public Domain *Contributors*: Andrewrabbott (talk). Original uploader was Andrewrabbott at en.wikipedia

Image:Jungfraubahn close-up both pantographs.jpg *Source*: http://en.wikipedia.org/w/index.php?title=File:Jungfraubahn_close-up_both_pantographs.jpg *License*: Public Domain *Contributors*: Gürbetaler, LennartBolks, Spsmiler, ZorkNika, 1 anonymous edits

Image:Rowanzug He 2-2 6 Antrieb 02 08.jpg *Source*: http://en.wikipedia.org/w/index.php?title=File:Rowanzug_He_2-2_6_Antrieb_02_08.jpg *License*: Creative Commons Attribution-Sharealike 3.0 *Contributors*: User:Chriusha

Image:20080927Y540 Strub.jpg *Source*: http://en.wikipedia.org/w/index.php?title=File:20080927Y540_Strub.jpg *License*: unknown *Contributors*: User:Gürbetaler

file:Mh spb mit kuh.jpeg *Source*: http://en.wikipedia.org/w/index.php?title=File:Mh_spb_mit_kuh.jpeg *License*: GNU Free Documentation License *Contributors*: User:LosHawlos

Image:SPB at Schynige Platte Station.jpg *Source*: http://en.wikipedia.org/w/index.php?title=File:SPB_at_Schynige_Platte_Station.jpg *License*: Creative Commons Attribution-Sharealike 2.5 *Contributors*: User:SalomonCeb

Image:20070816S715 14.jpg *Source*: http://en.wikipedia.org/w/index.php?title=File:20070816S715_14.jpg *License*: unknown *Contributors*: User:Gürbetaler

File:20080629Y407a SPB 11 Wilderswil.jpg *Source*: http://en.wikipedia.org/w/index.php?title=File:20080629Y407a_SPB_11_Wilderswil.jpg *License*: unknown *Contributors*: User:Gürbetaler

Image:20080629Y405a SPB 11.jpg *Source*: http://en.wikipedia.org/w/index.php?title=File:20080629Y405a_SPB_11.jpg *License*: unknown *Contributors*: User:Gürbetaler

Image:20080629Y522a 12.jpg *Source*: http://en.wikipedia.org/w/index.php?title=File:20080629Y522a_12.jpg *License*: unknown *Contributors*: User:Gürbetaler

File:20080629Y509a SPB 13 Matten.jpg *Source*: http://en.wikipedia.org/w/index.php?title=File:20080629Y509a_SPB_13_Matten.jpg *License*: unknown *Contributors*: User:Gürbetaler

Image:20080629Y484a SPB 13 Matten.jpg *Source*: http://en.wikipedia.org/w/index.php?title=File:20080629Y484a_SPB_13_Matten.jpg *License*: unknown *Contributors*: User:Gürbetaler

Image:20070816S715a SPB 14.jpg *Source*: http://en.wikipedia.org/w/index.php?title=File:20070816S715a_SPB_14.jpg *License*: unknown *Contributors*: User:Gürbetaler

File:20080629Y352a SPB 16 Anemone.jpg *Source*: http://en.wikipedia.org/w/index.php?title=File:20080629Y352a_SPB_16_Anemone.jpg *License*: unknown *Contributors*: User:Gürbetaler

Image:20080629Y429a SPB 16 Anemone.jpg *Source*: http://en.wikipedia.org/w/index.php?title=File:20080629Y429a_SPB_16_Anemone.jpg *License*: unknown *Contributors*: User:Gürbetaler

File:20080629Y363a SPB 18 Guendlischwand.jpg *Source*: http://en.wikipedia.org/w/index.php?title=File:20080629Y363a_SPB_18_Guendlischwand.jpg *License*: unknown *Contributors*: User:Gürbetaler

Image:20080629Y365a 18.jpg *Source*: http://en.wikipedia.org/w/index.php?title=File:20080629Y365a_18.jpg *License*: unknown *Contributors*: User:Gürbetaler

File:20080629Y539a SPB 19 Flühblume.jpg *Source*: http://en.wikipedia.org/w/index.php?title=File:20080629Y539a_SPB_19_Flühblume.jpg *License*: unknown *Contributors*: User:Gürbetaler

Image:20080629Y496a SPB 19 Flühblume.jpg *Source*: http://en.wikipedia.org/w/index.php?title=File:20080629Y496a_SPB_19_Flühblume.jpg *License*: unknown *Contributors*: User:Gürbetaler

File:20080629Y427a SPB 20 Gsteigwiler.jpg *Source*: http://en.wikipedia.org/w/index.php?title=File:20080629Y427a_SPB_20_Gsteigwiler.jpg *License*: unknown *Contributors*: User:Gürbetaler

Image:20080629Y426a SPB 20 Gsteigwiler.jpg *Source*: http://en.wikipedia.org/w/index.php?title=File:20080629Y426a_SPB_20_Gsteigwiler.jpg *License*: unknown *Contributors*: User:Gürbetaler

File:20080629Y546a SPB 61 Enzian.jpg *Source*: http://en.wikipedia.org/w/index.php?title=File:20080629Y546a_SPB_61_Enzian.jpg *License*: unknown *Contributors*: User:Gürbetaler

Image:20080629Y546a SPB 61.jpg *Source*: http://en.wikipedia.org/w/index.php?title=File:20080629Y546a_SPB_61.jpg *License*: unknown *Contributors*: User:Gürbetaler

File:20080629Y304a SPB 62 Alpenrose.jpg *Source*: http://en.wikipedia.org/w/index.php?title=File:20080629Y304a_SPB_62_Alpenrose.jpg *License*: unknown *Contributors*: User:Gürbetaler

Image:20080629Y328a SPB 62 Alpenrose.jpg *Source*: http://en.wikipedia.org/w/index.php?title=File:20080629Y328a_SPB_62_Alpenrose.jpg *License*: unknown *Contributors*: User:Gürbetaler

Image:20080629Y568a SPB 63.jpg *Source*: http://en.wikipedia.org/w/index.php?title=File:20080629Y568a_SPB_63.jpg *License*: unknown *Contributors*: User:Gürbetaler

file:Mh blm winteregg.jpeg *Source*: http://en.wikipedia.org/w/index.php?title=File:Mh_blm_winteregg.jpeg *License*: GNU Free Documentation License *Contributors*: Gürbetaler, LosHawlos, ZorkNika

Image:Mh blm lauterbr.jpeg *Source*: http://en.wikipedia.org/w/index.php?title=File:Mh_blm_lauterbr.jpeg *License*: GNU Free Documentation License *Contributors*: Kneiphof, LosHawlos, Pechristener, ZorkNika

Image:MurrenStationapproach.JPG *Source*: http://en.wikipedia.org/w/index.php?title=File:MurrenStationapproach.JPG *License*: Public Domain *Contributors*: Original uploader was Andrewrabbott at en.wikipedia

License

Creative Commons Attribution-ShareAlike 3.0 Unported - Deed

This is a human-readable summary of the Creative Commons Attribution ShareAlike 3.0 Unported License (http://en.wikipedia.org/wiki/Wikipedia:Text_of_Creative_Commons_Attribution-ShareAlike_3.0_Unported_License)
You are free:

- **to Share**—to copy, distribute and transmit the work, and
- **to Remix**—to adapt the work

Under the following conditions:

- **Attribution**—You must attribute the work in the manner specified by the author or licensor (but not in any way that suggests that they endorse you or your use of the work.)
- **Share Alike**—If you alter, transform, or build upon this work, you may distribute the resulting work only under the same, similar or a compatible license.

With the understanding that:

- **Waiver**—Any of the above conditions can be waived if you get permission from the copyright holder.
- **Other Rights**—In no way are any of the following rights affected by the license:
  - your fair dealing or fair use rights;
  - the author's moral rights; and
  - rights other persons may have either in the work itself or in how the work is used, such as publicity or privacy rights.
- **Notice**—For any reuse or distribution, you must make clear to others the license terms of this work. The best way to do that is with a link to http://creativecommons.org/licenses/by-sa/3.0/

GNU Free Documentation License

As of July 15, 2009 Wikipedia has moved to a dual-licensing system that supersedes the previous GFDL only licensing. In short, this means that text licensed under the GFDL can no longer be imported to Wikipedia. Additionally, text contributed after that date can not be exported under the GFDL license. See Wikipedia:Licensing update for further information.

Version 1.3, 3 November 2008 Copyright (C) 2000, 2001, 2002, 2007, 2008 Free Software Foundation, Inc. <http://fsf.org/>
Everyone is permitted to copy and distribute verbatim copies of this license document, but changing it is not allowed.

0. PREAMBLE

The purpose of this License is to make a manual, textbook, or other functional and useful document "free" in the sense of freedom: to assure everyone the effective freedom to copy and redistribute it, with or without modifying it, either commercially or noncommercially. Secondarily, this License preserves for the author and publisher a way to get credit for their work, while not being considered responsible for modifications made by others.
This License is a kind of "copyleft", which means that derivative works of the document must themselves be free in the same sense. It complements the GNU General Public License, which is a copyleft license designed for free software.
We have designed this License in order to use it for manuals for free software, because free software needs free documentation: a free program should come with manuals providing the same freedoms that the software does. But this License is not limited to software manuals; it can be used for any textual work, regardless of subject matter or whether it is published as a printed book. We recommend this License principally for works whose purpose is instruction or reference.

1. APPLICABILITY AND DEFINITIONS

This License applies to any manual or other work, in any medium, that contains a notice placed by the copyright holder saying it can be distributed under the terms of this License. Such a notice grants a world-wide, royalty-free license, unlimited in duration, to use that work under the conditions stated herein. The "Document", below, refers to any such manual or work. Any member of the public is a licensee, and is addressed as "you". You accept the license if you copy, modify or distribute the work in a way requiring permission under copyright law.
A "Modified Version" of the Document means any work containing the Document or a portion of it, either copied verbatim, or with modifications and/or translated into another language.
A "Secondary Section" is a named appendix or a front-matter section of the Document that deals exclusively with the relationship of the publishers or authors of the Document to the Document's overall subject (or to related matters) and contains nothing that could fall directly within that overall subject. (Thus, if the Document is in part a textbook of mathematics, a Secondary Section may not explain any mathematics.) The relationship could be a matter of historical connection with the subject or with related matters, or of legal, commercial, philosophical, ethical or political position regarding them.
The "Invariant Sections" are certain Secondary Sections whose titles are designated, as being those of Invariant Sections, in the notice that says that the Document is released under this License. If a section does not fit the above definition of Secondary then it is not allowed to be designated as Invariant. The Document may contain zero Invariant Sections. If the Document does not identify any Invariant Sections then there are none.
The "Cover Texts" are certain short passages of text that are listed, as Front-Cover Texts or Back-Cover Texts, in the notice that says that the Document is released under this License. A Front-Cover Text may be at most 5 words, and a Back-Cover Text may be at most 25 words.
A "Transparent" copy of the Document means a machine-readable copy, represented in a format whose specification is available to the general public, that is suitable for revising the document straightforwardly with generic text editors or (for images composed of pixels) generic paint programs or (for drawings) some widely available drawing editor, and that is suitable for input to text formatters or for automatic translation to a variety of formats suitable for input to text formatters. A copy made in an otherwise Transparent file format whose markup, or absence of markup, has been arranged to thwart or discourage subsequent modification by readers is not Transparent. An image format is not Transparent if used for any substantial amount of text. A copy that is not "Transparent" is called "Opaque".
Examples of suitable formats for Transparent copies include plain ASCII without markup, Texinfo input format, LaTeX input format, SGML or XML using a publicly available DTD, and standard-conforming simple HTML, PostScript or PDF designed for human modification. Examples of transparent image formats include PNG, XCF and JPG. Opaque formats include proprietary formats that can be read and edited only by proprietary word processors, SGML or XML for which the DTD and/or processing tools are not generally available, and the machine-generated HTML, PostScript or PDF produced by some word processors for output purposes only.
The "Title Page" means, for a printed book, the title page itself, plus such following pages as are needed to hold, legibly, the material this License requires to appear in the title page. For works in formats which do not have any title page as such, "Title Page" means the text near the most prominent appearance of the work's title, preceding the beginning of the body of the text.
The "publisher" means any person or entity that distributes copies of the Document to the public.
A section "Entitled XYZ" means a named subunit of the Document whose title either is precisely XYZ or contains XYZ in parentheses following text that translates XYZ in another language. (Here XYZ stands for a specific section name mentioned below, such as "Acknowledgements", "Dedications", "Endorsements", or "History".) To "Preserve the Title" of such a section when you modify the Document means that it remains a section "Entitled XYZ" according to this definition.
The Document may include Warranty Disclaimers next to the notice which states that this License applies to the Document. These Warranty Disclaimers are considered to be included by reference in this License, but only as regards disclaiming warranties: any other implication that these Warranty Disclaimers may have is void and has no effect on the meaning of this License.

2. VERBATIM COPYING

You may copy and distribute the Document in any medium, either commercially or noncommercially, provided that this License, the copyright notices, and the license notice saying this License applies to the Document are reproduced in all copies, and that you add no other conditions whatsoever to those of this License. You may not use technical measures to obstruct or control the reading or further copying of the copies you make or distribute. However, you may accept compensation in exchange for copies. If you distribute a large enough number of copies you must also follow the conditions in section 3.
You may also lend copies, under the same conditions stated above, and you may publicly display copies.

3. COPYING IN QUANTITY

If you publish printed copies (or copies in media that commonly have printed covers) of the Document, numbering more than 100, and the Document's license notice requires Cover Texts, you must enclose the copies in covers that carry, clearly and legibly, all these Cover Texts: Front-Cover Texts on the front cover, and Back-Cover Texts on the back cover. Both covers must also clearly and legibly identify you as the publisher of these copies. The front cover must present the full title with all words of the title equally prominent and visible. You may add other material on the covers in addition. Copying with changes limited to the covers, as long as they preserve the title of the Document and satisfy these conditions, can be treated as verbatim copying in other respects.
If the required texts for either cover are too voluminous to fit legibly, you should put the first ones listed (as many as fit reasonably) on the actual cover, and continue the rest onto adjacent pages.
If you publish or distribute Opaque copies of the Document numbering more than 100, you must either include a machine-readable Transparent copy along with each Opaque copy, or state in or with each Opaque copy a computer-network location from which the general network-using public has access to download using public-standard network protocols a complete Transparent copy of the Document, free of added material. If you use the latter option, you must take reasonably prudent steps, when you begin distribution of Opaque copies in quantity, to ensure that this Transparent copy will remain thus accessible at the stated location until at least one year after the last time you distribute an Opaque copy (directly or through your agents or retailers) of that edition to the public.
It is requested, but not required, that you contact the authors of the Document well before redistributing any large number of copies, to give them a chance to provide you with an updated version of the Document.

4. MODIFICATIONS

You may copy and distribute a Modified Version of the Document under the conditions of sections 2 and 3 above, provided that you release the Modified Version under precisely this License, with the Modified Version filling the role of the Document, thus licensing distribution and modification of the Modified Version to whoever possesses a copy of it. In addition, you must do these things in the Modified Version:

A. Use in the Title Page (and on the covers, if any) a title distinct from that of the Document, and from those of previous versions (which should, if there were any, be listed in the History section of the Document). You may use the same title as a previous version if the original publisher of that version gives permission.
B. List on the Title Page, as authors, one or more persons or entities responsible for authorship of the modifications in the Modified Version, together with at least five of the principal authors of the Document (all of its principal authors, if it has fewer than five), unless they release you from this requirement.
C. State on the Title page the name of the publisher of the Modified Version, as the publisher.
D. Preserve all the copyright notices of the Document.
E. Add an appropriate copyright notice for your modifications adjacent to the other copyright notices.
F. Include, immediately after the copyright notices, a license notice giving the public permission to use the Modified Version under the terms of this License, in the form shown in the Addendum below.
G. Preserve in that license notice the full lists of Invariant Sections and required Cover Texts given in the Document's license notice.
H. Include an unaltered copy of this License.
I. Preserve the section Entitled "History", Preserve its Title, and add to it an item stating at least the title, year, new authors, and publisher of the Modified Version as given on the Title Page. If there is no section Entitled "History" in the Document, create one stating the title, year, authors, and publisher of the Document as given on its Title Page, then add an item describing the Modified Version as stated in the previous sentence.
J. Preserve the network location, if any, given in the Document for public access to a Transparent copy of the Document, and likewise the network locations given in the Document for previous versions it was based on. These may be placed in the "History" section. You may omit a network location for a work that was published at least four years before the Document itself, or if the original publisher of the version it refers to gives permission.
K. For any section Entitled "Acknowledgements" or "Dedications", Preserve the Title of the section, and preserve in the section all the substance and tone of each of the contributor acknowledgements and/or dedications given therein.
L. Preserve all the Invariant Sections of the Document, unaltered in their text and in their titles. Section numbers or the equivalent are not considered part of the section titles.
M. Delete any section Entitled "Endorsements". Such a section may not be included in the Modified version.
N. Do not retitle any existing section to be Entitled "Endorsements" or to conflict in title with any Invariant Section.
O. Preserve any Warranty Disclaimers.

If the Modified Version includes new front-matter sections or appendices that qualify as Secondary Sections and contain no material copied from the Document, you may at your option designate some or all of these sections as invariant. To do this, add their titles to the list of Invariant Sections in the Modified Version's license notice. These titles must be distinct from any other section titles.
You may add a section Entitled "Endorsements", provided it contains nothing but endorsements of your Modified Version by various parties—for example, statements of peer review or that the text has been approved by an organization as the authoritative definition of a standard.
You may add a passage of up to five words as a Front-Cover Text, and a passage of up to 25 words as a Back-Cover Text, to the end of the list of Cover Texts in the Modified Version. Only one passage of Front-Cover Text and one of Back-Cover Text may be added by (or through arrangements made by) any one entity. If the Document already includes a cover text for the same cover, previously added by you or by arrangement made by the same entity you are acting on behalf of, you may not add another; but you may replace the old one, on explicit permission from the previous publisher that added the old one.
The author(s) and publisher(s) of the Document do not by this License give permission to use their names for publicity for or to assert or imply endorsement of any Modified Version.

5. COMBINING DOCUMENTS

You may combine the Document with other documents released under this License, under the terms defined in section 4 above for modified versions, provided that you include in the combination all of the Invariant Sections of all of the original documents, unmodified, and list them all as Invariant Sections of your combined work in its license notice, and that you preserve all their Warranty Disclaimers.
The combined work need only contain one copy of this License, and multiple identical Invariant Sections may be replaced with a single copy. If there are multiple Invariant Sections with the same name but different contents, make the title of each such section unique by adding at the end of it, in parentheses, the name of the original author or publisher of that section if known, or else a unique number. Make the same adjustment to the section titles in the list of Invariant Sections in the license notice of the combined work.
In the combination, you must combine any sections Entitled "History" in the various original documents, forming one section Entitled "History"; likewise combine any sections Entitled "Acknowledgements", and any sections Entitled "Dedications". You must delete all sections Entitled "Endorsements".

6. COLLECTIONS OF DOCUMENTS

You may make a collection consisting of the Document and other documents released under this License, and replace the individual copies of this License in the various documents with a single copy that is included in the collection, provided that you follow the rules of this License for verbatim copying of each of the documents in all other respects.
You may extract a single document from such a collection, and distribute it individually under this License, provided you insert a copy of this License into the extracted document, and follow this License in all other respects regarding verbatim copying of that document.

7. AGGREGATION WITH INDEPENDENT WORKS

A compilation of the Document or its derivatives with other separate and independent documents or works, in or on a volume of a storage or distribution medium, is called an "aggregate" if the copyright resulting from the compilation is not used to limit the legal rights of the compilation's users beyond what the individual works permit. When the Document is included in an aggregate, this License does not apply to the other works in the aggregate which are not themselves derivative works of the Document.

If the Cover Text requirement of section 3 is applicable to these copies of the Document, then if the Document is less than one half of the entire aggregate, the Document's Cover Texts may be placed on covers that bracket the Document within the aggregate, or the electronic equivalent of covers if the Document is in electronic form. Otherwise they must appear on printed covers that bracket the whole aggregate.

8. TRANSLATION

Translation is considered a kind of modification, so you may distribute translations of the Document under the terms of section 4. Replacing Invariant Sections with translations requires special permission from their copyright holders, but you may include translations of some or all Invariant Sections in addition to the original versions of these Invariant Sections. You may include a translation of this License, and all the license notices in the Document, and any Warranty Disclaimers, provided that you also include the original English version of this License and the original versions of those notices and disclaimers. In case of a disagreement between the translation and the original version of this License or a notice or disclaimer, the original version will prevail.

If a section in the Document is Entitled "Acknowledgements", "Dedications", or "History", the requirement (section 4) to Preserve its Title (section 1) will typically require changing the actual title.

9. TERMINATION

You may not copy, modify, sublicense, or distribute the Document except as expressly provided under this License. Any attempt otherwise to copy, modify, sublicense, or distribute it is void, and will automatically terminate your rights under this License.

However, if you cease all violation of this License, then your license from a particular copyright holder is reinstated (a) provisionally, unless and until the copyright holder explicitly and finally terminates your license, and (b) permanently, if the copyright holder fails to notify you of the violation by some reasonable means prior to 60 days after the cessation.

Moreover, your license from a particular copyright holder is reinstated permanently if the copyright holder notifies you of the violation by some reasonable means, this is the first time you have received notice of violation of this License (for any work) from that copyright holder, and you cure the violation prior to 30 days after your receipt of the notice.

Termination of your rights under this section does not terminate the licenses of parties who have received copies or rights from you under this License. If your rights have been terminated and not permanently reinstated, receipt of a copy of some or all of the same material does not give you any rights to use it.

10. FUTURE REVISIONS OF THIS LICENSE

The Free Software Foundation may publish new, revised versions of the GNU Free Documentation License from time to time. Such new versions will be similar in spirit to the present version, but may differ in detail to address new problems or concerns. See http://www.gnu.org/copyleft/.

Each version of the License is given a distinguishing version number. If the Document specifies that a particular numbered version of this License "or any later version" applies to it, you have the option of following the terms and conditions either of that specified version or of any later version that has been published (not as a draft) by the Free Software Foundation. If the Document does not specify a version number of this License, you may choose any version ever published (not as a draft) by the Free Software Foundation. If the Document specifies that a proxy can decide which future versions of this License can be used, that proxy's public statement of acceptance of a version permanently authorizes you to choose that version for the Document.

11. RELICENSING

"Massive Multiauthor Collaboration Site" (or "MMC Site") means any World Wide Web server that publishes copyrightable works and also provides prominent facilities for anybody to edit those works. A public wiki that anybody can edit is an example of such a server. A "Massive Multiauthor Collaboration" (or "MMC") contained in the site means any set of copyrightable works thus published on the MMC site.

"CC-BY-SA" means the Creative Commons Attribution-Share Alike 3.0 license published by Creative Commons Corporation, a not-for-profit corporation with a principal place of business in San Francisco, California, as well as future copyleft versions of that license published by that same organization.

"Incorporate" means to publish or republish a Document, in whole or in part, as part of another Document.

An MMC is "eligible for relicensing" if it is licensed under this License, and if all works that were first published under this License somewhere other than this MMC, and subsequently incorporated in whole or in part into the MMC, (1) had no cover texts or invariant sections, and (2) were thus incorporated prior to November 1, 2008.

The operator of an MMC Site may republish an MMC contained in the site under CC-BY-SA on the same site at any time before August 1, 2009, provided the MMC is eligible for relicensing.

How to use this License for your documents

To use this License in a document you have written, include a copy of the License in the document and put the following copyright and license notices just after the title page:

> Copyright (c) YEAR YOUR NAME.
>
> Permission is granted to copy, distribute and/or modify this document
>
> under the terms of the GNU Free Documentation License, Version 1.3
>
> or any later version published by the Free Software Foundation;
>
> with no Invariant Sections, no Front-Cover Texts, and no Back-Cover Texts.
>
> A copy of the license is included in the section entitled "GNU
>
> Free Documentation License".

If you have Invariant Sections, Front-Cover Texts and Back-Cover Texts, replace the "with...Texts." line with this:

> with the Invariant Sections being LIST THEIR TITLES, with the
>
> Front-Cover Texts being LIST, and with the Back-Cover Texts being LIST.

If you have Invariant Sections without Cover Texts, or some other combination of the three, merge those two alternatives to suit the situation.

If your document contains nontrivial examples of program code, we recommend releasing these examples in parallel under your choice of free software license, such as the GNU General Public License, to permit their use in free software.

GNU Free Documentation License Version 1.2, November 2002 Copyright (C) 2000,2001,2002 Free Software Foundation, Inc. 59 Temple Place, Suite 330, Boston, MA 02111-1307 USA Everyone is permitted to copy and distribute verbatim copies of this license document, but changing it is not allowed.

0. PREAMBLE

The purpose of this License is to make a manual, textbook, or other functional and useful document "free" in the sense of freedom: to assure everyone the effective freedom to copy and redistribute it, with or without modifying it, either commercially or noncommercially. Secondarily, this License preserves for the author and publisher a way to get credit for their work, while not being considered responsible for modifications made by others. This License is a kind of "copyleft", which means that derivative works of the document must themselves be free in the same sense. It complements the GNU General Public License, which is a copyleft license designed for free software. We have designed this License in order to use it for manuals for free software, because free software needs free documentation: a free program should come with manuals providing the same freedoms that the software does. But this License is not limited to software manuals; it can be used for any textual work, regardless of subject matter or whether it is published as a printed book. We recommend this License principally for works whose purpose is instruction or reference.

1. APPLICABILITY AND DEFINITIONS

This License applies to any manual or other work, in any medium, that contains a notice placed by the copyright holder saying it can be distributed under the terms of this License. Such a notice grants a world-wide, royalty-free license, unlimited in duration, to use that work under the conditions stated herein. The "Document", below, refers to any such manual or work. Any member of the public is a licensee, and is addressed as "you". You accept the license if you copy, modify or distribute the work in a way requiring permission under copyright law. A "Modified Version" of the Document means any work containing the Document or a portion of it, either copied verbatim, or with modifications and/or translated into another language. A "Secondary Section" is a named appendix or a front-matter section of the Document that deals exclusively with the relationship of the publishers or authors of the Document to the Document's overall subject (or to related matters) and contains nothing that could fall directly within that overall subject. (Thus, if the Document is in part a textbook of mathematics, a Secondary Section may not explain any mathematics.) The relationship could be a matter of historical connection with the subject or with related matters, or of legal, commercial, philosophical, ethical or political position regarding them. The "Invariant Sections" are certain Secondary Sections whose titles are designated, as being those of Invariant Sections, in the notice that says that the Document is released under this License. If a section does not fit the above definition of Secondary then it is not allowed to be designated as Invariant. The Document may contain zero Invariant Sections. If the Document does not identify any Invariant Sections then there are none. The "Cover Texts" are certain short passages of text that are listed, as Front-Cover Texts or Back-Cover Texts, in the notice that says that the Document is released under this License. A Front-Cover Text may be at most 5 words, and a Back-Cover Text may be at most 25 words. A "Transparent" copy of the Document means a machine-readable copy, represented in a format whose specification is available to the general public, that is suitable for revising the document straightforwardly with generic text editors or (for images composed of pixels) generic paint programs or (for drawings) some widely available drawing editor, and that is suitable for input to text formatters or for automatic translation to a variety of formats suitable for input to text formatters. A copy made in an otherwise Transparent file format whose markup, or absence of markup, has been arranged to thwart or discourage subsequent modification by readers is not Transparent. An image format is not Transparent if used for any substantial amount of text. A copy that is not "Transparent" is called "Opaque". Examples of suitable formats for Transparent copies include plain ASCII without markup, Texinfo input format, LaTeX input format, SGML or XML using a publicly available DTD, and standard-conforming simple HTML, PostScript or PDF designed for human modification. Examples of transparent image formats include PNG, XCF and JPG. Opaque formats include proprietary formats that can be read and edited only by proprietary word processors, SGML or XML for which the DTD and/or processing tools are not generally available, and the machine-generated HTML, PostScript or PDF produced by some word processors for output purposes only. The "Title Page" means, for a printed book, the title page itself, plus such following pages as are needed to hold, legibly, the material this License requires to appear in the title page. For works in formats which do not have any title page as such, "Title Page" means the text near the most prominent appearance of the work's title, preceding the beginning of the body of the text. A section "Entitled XYZ" means a named subunit of the Document whose title either is precisely XYZ or contains XYZ in parentheses following text that translates XYZ in another language. (Here XYZ stands for a specific section name mentioned below, such as "Acknowledgements", "Dedications", "Endorsements", or "History".) To "Preserve the Title" of such a section when you modify the Document means that it remains a section "Entitled XYZ" according to this definition. The Document may include Warranty Disclaimers next to the notice which states that this License applies to the Document. These Warranty Disclaimers are considered to be included by reference in this License, but only as regards disclaiming warranties: any other implication that these Warranty Disclaimers may have is void and has no effect on the meaning of this License.

2. VERBATIM COPYING

You may copy and distribute the Document in any medium, either commercially or noncommercially, provided that this License, the copyright notices, and the license notice saying this License applies to the Document are reproduced in all copies, and that you add no other conditions whatsoever to those of this License. You may not use technical measures to obstruct or control the reading or further copying of the copies you make or distribute. However, you may accept compensation in exchange for copies. If you distribute a large enough number of copies you must also follow the conditions in section 3. You may also lend copies, under the same conditions stated above, and you may publicly display copies.

3. COPYING IN QUANTITY

If you publish printed copies (or copies in media that commonly have printed covers) of the Document, numbering more than 100, and the Document's license notice requires Cover Texts, you must enclose the copies in covers that carry, clearly and legibly, all these Cover Texts: Front-Cover Texts on the front cover, and Back-Cover Texts on the back cover. Both covers must also clearly and legibly identify you as the publisher of these copies. The front cover must present the full title with all words of the title equally prominent and visible. You may add other material on the covers in addition. Copying with changes limited to the covers, as long as they preserve the title of the Document and satisfy these conditions, can be treated as verbatim copying in other respects. If the required texts for either cover are too voluminous to fit legibly, you should put the first ones listed (as many as fit reasonably) on the actual cover, and continue the rest onto adjacent pages. If you publish or distribute Opaque copies of the Document numbering more than 100, you must either include a machine-readable Transparent copy along with each Opaque copy, or state in or with each Opaque copy a computer-network location from which the general network-using public has access to download using public-standard network protocols a complete Transparent copy of the Document, free of added material. If you use the latter option, you must take reasonably prudent steps, when you begin distribution of Opaque copies in quantity, to ensure that this Transparent copy will remain thus accessible at the stated location until at least one year after the last time you distribute an Opaque copy (directly or through your agents or retailers) of that edition to the public. It is requested, but not required, that you contact the authors of the Document well before redistributing any large number of copies, to give them a chance to provide you with an updated version of the Document.

4. MODIFICATIONS

You may copy and distribute a Modified Version of the Document under the conditions of sections 2 and 3 above, provided that you release the Modified Version under precisely this License, with the Modified Version filling the role of the Document, thus licensing distribution and modification of the Modified Version to whoever possesses a copy of it. In addition, you must do these things in the Modified Version: A. Use in the Title Page (and on the covers, if any) a title distinct from that of the Document, and from those of previous versions (which should, if there were any, be listed in the History section of the Document). You may use the same title as a previous version if the original publisher of that version gives permission. B. List on the Title Page, as authors, one or more persons or entities responsible for authorship of the modifications in the Modified Version, together with at least five of the principal authors of the Document (all of its principal authors, if it has fewer than five), unless they release you from this requirement. C. State on the Title page the name of the publisher of the Modified Version, as the publisher. D. Preserve all the copyright notices of the Document. E. Add an appropriate copyright notice for your modifications adjacent to the other copyright notices. F. Include, immediately after the copyright notices, a license notice giving the public permission to use the Modified Version under the terms of this License, in the form shown in the Addendum below. G. Preserve in that license notice the full lists of Invariant Sections and required Cover Texts given in the Document's license notice. H. Include an unaltered copy of this License. I. Preserve the section Entitled "History", Preserve its Title, and add to it an item stating at least the title, year, new authors, and publisher of the Modified Version as given on the Title Page. If there is no section Entitled "History" in the Document, create one stating the title, year, authors, and publisher of the Document as given on its Title Page, then add an item describing the Modified Version as stated in the previous sentence. J. Preserve the network location, if any, given in the Document for public access to a Transparent copy of the Document, and likewise the network locations given in the Document for previous versions it was based on. These may be placed in the "History" section. You may omit a network location for a work that was published at least four years before the Document itself, or if the original publisher of the version it refers to gives permission. K. For any section Entitled "Acknowledgements" or "Dedications", Preserve the Title of the section, and preserve in the section all the substance and tone of each of the contributor acknowledgements and/or dedications given therein. L. Preserve all the Invariant Sections of the Document, unaltered in their text and in their titles. Section numbers or the equivalent are not considered part of the section titles. M. Delete any section Entitled "Endorsements". Such a section may not be included in the Modified Version. N. Do not retitle any existing section to be Entitled "Endorsements" or to conflict in title with any Invariant Section. O. Preserve any Warranty Disclaimers. If the Modified Version includes new front-matter sections or appendices that qualify as Secondary Sections and contain no material copied from the Document, you may at your option designate some or all of these sections as invariant. To do this, add their titles to the list of Invariant Sections in the Modified Version's license notice. These titles must be distinct from any other section titles. You may add a section Entitled "Endorsements", provided it contains nothing but endorsements of your Modified Version by various parties--for example, statements of peer review or that the text has been approved by an organization as the authoritative definition of a standard. You may add a passage of up to five words as a Front-Cover Text, and a passage of up to 25 words as a Back-Cover Text, to the end of the list of Cover Texts in the Modified Version. Only one passage of Front-Cover Text and one of Back-Cover Text may be added by (or through arrangements made by) any one entity. If the Document already includes a cover text for the same cover, previously added by you or by arrangement made by the same entity you are acting on behalf of, you may not add another; but you may replace the old one, on explicit permission from the previous publisher that added the old one. The author(s) and publisher(s) of the Document do not by this License give permission to use their names for publicity for or to assert or imply endorsement of any Modified Version.

5. COMBINING DOCUMENTS

You may combine the Document with other documents released under this License, under the terms defined in section 4 above for modified versions, provided that you include in the combination all of the Invariant Sections of all of the original documents, unmodified, and list them all as Invariant Sections of your combined work in its license notice, and that you preserve all their Warranty Disclaimers. The combined work need only contain one copy of this License, and multiple identical Invariant Sections may be replaced with a single copy. If there are multiple Invariant Sections with the same name but different contents, make the title of each such section unique by adding at the end of it, in parentheses, the name of the original author or publisher of that section if known, or else a unique number. Make the same adjustment to the section titles in the list of Invariant Sections in the license notice of the combined work. In the combination, you must combine any sections Entitled "History" in the various original documents, forming one section Entitled "History"; likewise combine any sections Entitled "Acknowledgements", and any sections Entitled "Dedications". You must delete all sections Entitled "Endorsements".

6. COLLECTIONS OF DOCUMENTS

You may make a collection consisting of the Document and other documents released under this License, and replace the individual copies of this License in the various documents with a single copy that is included in the collection, provided that you follow the rules of this License for verbatim copying of each of the documents in all other respects. You may extract a single document from such a collection, and distribute it individually under this License, provided you insert a copy of this License into the extracted document, and follow this License in all other respects regarding verbatim copying of that document.

7. AGGREGATION WITH INDEPENDENT WORKS

A compilation of the Document or its derivatives with other separate and independent documents or works, in or on a volume of a storage or distribution medium, is called an "aggregate" if the copyright resulting from the compilation is not used to limit the legal rights of the compilation's users beyond what the individual works permit. When the Document is included in an aggregate, this License does not apply to the other works in the aggregate which are not themselves derivative works of the Document. If the Cover Text requirement of section 3 is applicable to these copies of the Document, then if the Document is less than one half of the entire aggregate, the Document's Cover Texts may be placed on covers that bracket the Document within the aggregate, or the electronic equivalent of covers if the Document is in electronic form. Otherwise they must appear on printed covers that bracket the whole aggregate.

8. TRANSLATION

Translation is considered a kind of modification, so you may distribute translations of the Document under the terms of section 4. Replacing Invariant Sections with translations requires special permission from their copyright holders, but you may include translations of some or all Invariant Sections in addition to the original versions of these Invariant Sections. You may include a translation of this License, and all the license notices in the Document, and any Warranty Disclaimers, provided that you also include the original English version of this License and the original versions of those notices and disclaimers. In case of a disagreement between the translation and the original version of this License or a notice or disclaimer, the original version will prevail. If a section in the Document is Entitled "Acknowledgements", "Dedications", or "History", the requirement (section 4) to Preserve its Title (section 1) will typically require changing the actual title.

9. TERMINATION

You may not copy, modify, sublicense, or distribute the Document except as expressly provided for under this License. Any other attempt to copy, modify, sublicense or distribute the Document is void, and will automatically terminate your rights under this License. However, parties who have received copies, or rights, from you under this License will not have their licenses terminated so long as such parties remain in full compliance.

10. FUTURE REVISIONS OF THIS LICENSE

The Free Software Foundation may publish new, revised versions of the GNU Free Documentation License from time to time. Such new versions will be similar in spirit to the present version, but may differ in detail to address new problems or concerns. See http://www.gnu.org/copyleft/. Each version of the License is given a distinguishing version number. If the Document specifies that a particular numbered version of this License "or any later version" applies to it, you have the option of following the terms and conditions either of that specified version or of any later version that has been published (not as a draft) by the Free Software Foundation. If the Document does not specify a version number of this License, you may choose any version ever published (not as a draft) by the Free Software Foundation. ADDENDUM: How to use this License for your documents To use this License in a document you have written, include a copy of the License in the document and put the following copyright and license notices just after the title page: Copyright (c) YEAR YOUR NAME. Permission is granted to copy, distribute and/or modify this document under the terms of the GNU Free Documentation License, Version 1.2 or any later version published by the Free Software Foundation; with no Invariant Sections, no Front-Cover Texts, and no Back-Cover Texts. A copy of the license is included in the section entitled "GNU Free Documentation License". If you have Invariant Sections, Front-Cover Texts and Back-Cover Texts, replace the "with...Texts." line with this: with the Invariant Sections being LIST THEIR TITLES, with the Front-Cover Texts being LIST, and with the Back-Cover Texts being LIST. If you have Invariant Sections without Cover Texts, or some other combination of the three, merge those two alternatives to suit the situation. If your document contains nontrivial examples of program code, we recommend releasing these examples in parallel under your choice of free software license, such as the GNU General Public License, to permit their use in free software.

Printed by Books on Demand GmbH, Norderstedt / Germany